SpringerBriefs in Applied Sciences and Technology

Continuum Mechanics

Series Editors

Holm Altenbach, Institut für Mechanik, Lehrstuhl für Technische Mechanik, Otto von Guericke University Magdeburg, Magdeburg, Sachsen-Anhalt, Germany

Andreas Öchsner, Faculty of Mechanical Engineering, Esslingen University of Applied Sciences, Esslingen am Neckar, Germany

These SpringerBriefs publish concise summaries of cutting-edge research and practical applications on any subject of Continuum Mechanics and Generalized Continua, including the theory of elasticity, heat conduction, thermodynamics, electromagnetic continua, as well as applied mathematics.

SpringerBriefs in Continuum Mechanics are devoted to the publication of fundamentals and applications, presenting concise summaries of cutting-edge research and practical applications across a wide spectrum of fields. Featuring compact volumes of 50 to 125 pages, the series covers a range of content from professional to academic.

More information about this subseries at http://www.springer.com/series/10528

Christian B. Silbermann ·
Matthias Baitsch · Jörn Ihlemann

Introduction to Geometrically Nonlinear Continuum Dislocation Theory

FE Implementation and Application on Subgrain Formation in Cubic Single Crystals Under Large Strains

Springer

Christian B. Silbermann
Institute of Mechanics and Thermodynamics
Chemnitz University of Technology
Chemnitz, Germany

Matthias Baitsch
Bau- und Umweltingenieurwesen
Bochum University of Applied Sciences
Bochum, Germany

Jörn Ihlemann
Institute of Mechanics and Thermodynamics
Chemnitz University of Technology
Chemnitz, Germany

ISSN 2191-530X ISSN 2191-5318 (electronic)
SpringerBriefs in Applied Sciences and Technology
ISSN 2625-1329 ISSN 2625-1337 (electronic)
SpringerBriefs in Continuum Mechanics
ISBN 978-3-030-63695-1 ISBN 978-3-030-63696-8 (eBook)
https://doi.org/10.1007/978-3-030-63696-8

This Springer imprint is published by the registered company Springer Nature Switzerland AG
The registered company address is: Gewerbestrasse 11, 6330 Cham, Switzerland

«A good theoretical model of a complex system should be like a good caricature: it should emphasize those features which are most important and should downplay the inessential details.

Now the only snag with this advice is that one does not really know which are the inessential details until one has understood the phenomena under study.

Consequently, one should investigate a wide range of models and not stake one's life (or one's theoretical insight) on one particular model alone.»

Jakow I. Frenkel (1894–1952)[1]

---

[1]Fisher, M.E.: Scaling, universality and renormalization group theory. In: F.J.W. Hahne (ed.) Critical Phenomena, *Lecture Notes in Physics*, vol. 186, pp. 1–139. Springer, Berlin (1983).

Preface

This book deals with solid continuum mechanics applied to crystal plasticity on a mesoscopic length scale. The formation of dislocation structures and subgrain boundaries is decisive for the macroscopic properties of metallic materials. To simulate and predict this, Gurtin's established geometrically nonlinear mesoscopic theory of crystal plasticity with continuously distributed, geometrically necessary dislocations is adopted and specified. Here, the degrees of freedom essential for subgrain formation are available, while the number of phenomenological approaches and associated material parameters are kept as small as possible. With this continuum dislocation theory (CDT), it is proven that subgrain formation can be simulated. To this end, a minimalistic model of elastically and plastically anisotropic single crystal viscoplasticity is derived. Subsequently employing the finite element method (FEM), it is possible to follow the formation of subgrain boundaries during large plastic deformations of a crystallite. In addition, preparation and synthesis of algorithms for the numerical solution of the associated multi-field problem using the FEM are provided and aspects of thermodynamic consistency are discussed. Finally, proper experiments are suggested for the validation of the numerical results, and open problems are discussed for future research.

The intention of this book is threefold: Firstly, it presents a comprehensive introduction to single crystal plasticity with continuously distributed dislocations. Secondly, it provides a straightforward, extensive preparation for the implementation into FEM codes including solution algorithms in the context of a higher gradient theory. Thirdly, a simple simulation example introduces the characteristics of pattern forming systems and points out the numerical challenges involved in dealing with localization phenomena. Thus, the book may serve as a starting point for entering and contributing to the research in this interesting field of computational materials science.

This book is based on results previously published within Christian B. Silbermann's doctoral thesis (English title *On the modeling of dislocation- and deformation-induced plastic localization phenomena of metallic materials*, https://nbn-resolving.org/urn:nbn:de:bsz:ch1-qucosa2-359730).

Chemnitz, Germany
July 2020

Christian B. Silbermann
Matthias Baitsch
Jörn Ihlemann

Acknowledgements This research was supported by German Science Foundation (DFG) within the Collaborative Research Center SFB 692 HALS—High-strength aluminum-based lightweight materials for safety components. The authors are grateful to Prof. M. F.-X. Wagner, Prof. K. C. Le, M. Koster, Dr. Ralf Landgraf and Dr. Hendrik Donner for fruitful discussions. Moreover, the help of Hugh Wessel and the constructive criticism from the reviewers is acknowledged.

Contents

Notations, Symbols and Acronyms

Notations

$\underset{\sim}{e}_a$	Cartesian basis vectors, with $a \in \{x, y, z\}$
$\underset{\sim}{u}, \underset{\approx}{X}, \underset{\equiv}{V}, \underset{\equiv}{w}$	Tensor of first, second, third and fourth order
$\mathrm{fun}((\underset{\approx}{x}))$	Tensor-valued function of $\underset{\approx}{x}$, e. g. tensor exponential $\exp((\underset{\approx}{x}))$
$[u_a]$	Coeffizient matrix (column 3×1) of $\underset{\sim}{u}$ w.r.t. base $\underset{\sim}{e}_a$
$[X_{ab}]$	Coeffizient matrix (quadratic 3×3), from $\underset{\approx}{X}$ w. r. t. base $\underset{\sim}{e}_a$
$[M]$	General matrix (dimension follows from the context)
$\mathbf{x}$	Tuple of numbers
$\mathrm{d}*$	Differential (infinitesimal), e. g. infinitesimal volume element $\mathrm{d}V$
$\Delta*$	Perturbation/difference (finite)
$\delta*, \delta_{ab}$	Variation in the sense of variational calculus, Kronecker symbol
$G(*), G$	Shape function in the FEM, constant shear modulus
$\mathcal{K}$	Continuum mechanical configuration
$\tilde{*}, \hat{*}$	Accentuation as a quantity of the reference or some intermediate configuration
$\mathrm{\pi}$, π	Ratio of the circumference of a circle to its diameter, variable for the micro traction

Scalars

ℓ, u	Internal length scale, specific internal energy
$V, \widetilde{V}$	Volume in the current and reference configurations
$\varrho, \tilde{\varrho}$	Mass density in the current and reference configurations
ρ	Dislocation density (piercing points per area)
$\mathscr{r}$, r	Dimensionless dislocation density, residual of a zero form
p	Power density (power per volume in the current configuration)

$\phi, \tilde{\phi}$ Free energy density (energy per volume in the current and reference configurations)
$\psi, \tilde{\psi}$ Specific free energy (energy per mass in the current and reference configurations)
ζ, $\varkappa$ Specific entropy, entropy density (in the current configuration)
d Dissipation density (power per volume in the current configuration)
θ Thermodynamic temperature
φ Slip system orientation (angle)
υ Velocity, and $\underline{v}$ as a velocity vector
ν Slip rate
κ, $\varkappa$ Effective resolved shear stress, elastic constant
ϑ Angle of some plane lattice rotation

Vectors

$\underline{r}$ Placement vector
$\underline{u}$ Displacement vector
$\underline{b}$ Burgers vector, and $\underline{b}_r$ as resultant Burgers vector of a dislocation ensemble
$\underline{\alpha}, \underline{a}$ Dislocation density vector, crystal lattice vector
$\underline{s}, \underline{s}$ Stress vector, slip direction
$\underline{m}$ Slip plane normal
$\underline{n}$ Normal unit vector of some (cut) face
$\underline{t}$ Tangent vector
$\underline{q}$ Micro stress, i. e. higher order internal stress
$\underline{p}$ Distributed Peach-Koehler force

Tensors

$\underline{\underline{A}}$ Structure tensor
$\underline{\underline{\alpha}}$ Dislocation density tensor
$\underline{\underline{\beta}}$ Displacement gradient
$\underline{\underline{F}}$ Deformation (often called deformation gradient) with $j = I_3(\underline{\underline{F}})$
$\underline{\underline{V}}, \underline{\underline{\widetilde{U}}}$ Left- and right stretch tensor
$\underline{\underline{R}}$ Orthogonal tensor from the polar decomposition of the deformation
$\underline{\underline{b}}, \underline{\underline{\widetilde{C}}}$ Left- and right Cauchy Green tensor
$\underline{\underline{E}}$ Green strain tensor
$\underline{\underline{\varepsilon}}$ Almansi strain tensor
$\underline{\underline{L}}$ Velocity gradient

$\underline{\underline{D}}$ Strain rate tensor (distortion velocity)
$\underline{\underline{\Omega}}$ Spin tensor (rotation velocity)
$\underline{\underline{\sigma}}, \underline{\underline{\widetilde{T}}}$ Cauchy stress tensor, Second Piola Kirchhoff stress tensor
$\underline{\underline{S}}$ Slip system tensor

Index Variables

$a, b, c, \ldots$ Cartesian coordinates, $\in \{x, y, z\}$
J, K Cartesian coordinates, $\in \{\mathrm{I}, \mathrm{II}, \mathrm{III}\}$ for crystal or principal directions
L, M Number of FE shape function, $\in \{1, 2, \ldots, N\}$
i, j Number (e. g. of some slip system), $\in \{1, 2, \ldots\}$

Acronyms

BC Boundary condition
CDT Continuum dislocation theory
DOF Degree of freedom
FE Finite element
FEM Finite element method
GND Geometrically necessary dislocation
IC Initial condition
RSS Resolved shear stress
SS Slip system
SSD Statistically stored dislocations (redundant dislocations)

Chapter 1
Introduction

The mechanical behavior of crystals, especially metallic ones, strongly depends on the intrinsic defect structure. For a broad class of metals, the motion of dislocations carries the plastic distortion, and new macroscopic properties emerge from the collective self-organization of dislocations [1–6]. As a typical feature, homogeneous plastic distortion may become instable and localize in smaller domains of the material [7]. Understanding these localization phenomena is of great importance for many reasons: On the one hand, localized plastic distortion may be critical for the application of mechanical components. On the other hand, controlling and exploiting this self-organized (micro)structure formation (e. g. resulting in subgrain structures) may lead to desirable material properties. Moreover, considering localization and laminate formation, there are striking similarities between the microscopic occurrence in metallic materials and the macroscopic occurrence in geological materials (e. g. chevron folds as a collective buckling phenomenon of layered sedimentary rocks) [8, p. 34].

As typical structure formation processes of metallic crystals involve large dislocation ensembles, a *Continuum* Dislocation Theory (CDT) is advantageous. Currently there are many competing CDTs (for a comprehensive overview see e. g. [9, 10]). This multitude of theories and approaches seems to reflect Frenkel's remarkable statement on good models of complex systems (cf. p. v). However, most CDTs have in common that gradients of macroscopic plastic distortion are related to the presence of Geometrically Necessary Dislocations (GNDs).[1] As a key assumption, the dislocation tensor is considered as a thermodynamic state variable reflecting tensorial dislocation properties. For (large) dislocation ensembles the resulting dislocation tensor represents a homogenized quantity characterizing the effect of GNDs.

[1] In order to take into account *all* dislocations, some theories consider also Statistically Stored Dislocations (SSDs) [11], e. g. [12] or geometrically redundant dislocations [13].

C. B. Silbermann et al., *Introduction to Geometrically Nonlinear Continuum Dislocation Theory*, SpringerBriefs in Continuum Mechanics,
https://doi.org/10.1007/978-3-030-63696-8_1

The present book adopts the CDT proposed in [14] and developed further in a series of papers [15–17]. As a new feature, elastic anisotropy is introduced and the theory is specified for cubic primitive lattices with two slips systems. The main advantage of this specification is its clear and relatively simple structure. As an essential difference to the geometrically linear theory (cf. e. g. [18]), the deformation of the crystal lattice and thus of the slip systems is taken into account. The aim of this book is to obtain a minimalistic 2D model which contains the (nonlinear) effects essential for the dislocation-induced subgrain formation. These—so the thesis of the present work—are:

- elastic distortion of the crystals lattice (and the slip systems),
- plastic distortion as a consequence of slip processes in defined slip systems,
- energy dissipation due to plastic slip,
- energy storage due to elastic straining of the crystal lattice,
- energy storage due to elastic/plastic incompatibility.

The implementation of the theory in a finite element code finally allows the verification of this thesis by a systematic simulation study of a single crystal shearing.

The book is organized as follows: Chap. 2 reviews geometrically nonlinear crystal kinematics, including strain and GND measures. Based on that, Chap. 3 presents formulations for the free energy attributed to elastic strains and GNDs. Proposing a comprehensive and consistent thermodynamical framework, field equations are derived fulfilling the Clausius-Planck inequality. Thereby, energy dissipation due to dislocation motion is introduced in a thermodynamically consistent way. Chapter 4 summarizes special cases included in the presented theory. In addition, Sect. 4.4 adumbrates the geometrical linearization and discovers similarities and differences with regard to geometrically linear continuum dislocation theory [18]. Chapter 5 then derives a variational formulation of the theory based on the principle of virtual power [19]. Chapter 6 presents a comprehensive methodology for the numerical solution of the integral equations by means of the finite element method. For the special case of a continuously dislocated, plane, cubic primitive single crystal with two active slip systems the governing integral equations are transformed to matrix notation. Chapter 7 then presents numerical solutions of the corresponding initial boundary value problems for the case of plane shear deformations. To this end, an in-house simulation code using the Finite Element Method (FEM) is adopted. For the validation of the numerical results, Sect. 8.3 suggests some proper experiments. Finally, Chap. 8 discusses the presented specification of the theory, indicates existing challenges and problems and motivates directions for future research.

For the sake of compactness and clarity, symbolic tensor notation is preferred throughout this book. Tensors of nth order are denoted by a small or capital letter with n underscores, e. g. $\underline{\underline{U}}$ is a second-order tensor. The tensor product is denoted by $\otimes$, the cross product by $\times$ and the nth contraction of tensors is written symbolically with n dots $(\cdot)$. The coefficients with respect to a certain Cartesian coordinate system $\underline{e}_a$ with $a \in \{x, y, z\}$ are $U_{ab} = \underline{e}_a \cdot \underline{\underline{U}} \cdot \underline{e}_b$. The arrangement of second-order tensor coefficients in a quadratic matrix is denoted by $[U_{ab}]$. Spatial derivatives with respect to referential and current coordinates are expressed symbolically with the Nabla operators denoted by $\widetilde{\underline{\nabla}}$ and $\underline{\nabla}$, respectively. Wherever it facilitates reading, the derived operators are denoted as Grad, Div, Curl and grad, div, curl with respect to referential and current coordinates. An overview of the applied tensor calculus and analysis was given in [18].

References

1. Amodeo, R.J., Ghoniem, N.M.: Dislocation dynamics. I. A proposed methodology for deformation micromechanics. Phys.Rev. B **41**, 6958–6967 (1990)
2. Gregor, V.: Self-organization approach to cyclic microplasticity: a model of a persistent slip band. Int. J. Plast. **14**(1–3), 159–172 (1998)
3. Richeton, T., Dobron, P., Chmelik, F., Weiss, J., Louchet, F.: On the critical character of plasticity in metallic single crystals. Mater. Sci. Eng. A **424**(1–2), 190–195 (2006)
4. Chiu, Y., Veyssiere, P.: Dislocation self-organization under single slip straining and dipole properties. Mater. Sci. Eng. A **483–484**, 191–194 (2008)
5. Zahn, D., Tlatlik, H., Raabe, D.: Modeling of dislocation patterns of small- and high-angle grain boundaries in aluminum. Comput. Mater. Sci. **46**(2), 293–296 (2009)
6. Roters, F., Eisenlohr, P., Bieler, T.R., Raabe, D.: Crystal Plasticity Finite Element Methods in Materials Science and Engineering. Wiley, New York (2010)
7. Gao, Y.F., Larson, B.C., Lee, J.H., Nicola, L., Tischler, J.Z., Pharr, G.M.: Lattice rotation patterns and strain gradient effects in face-centered-cubic single crystals under spherical indentation. J. Appl. Mech. **82**(6), 061,007+ (2015)
8. Conti, S., Hackl, K. (eds.): Analysis and Computation of Microstructure in Finite Plasticity. Springer, Berlin (2015)
9. Sandfeld, S., Monavari, M., Zaiser, M.: From systems of discrete dislocations to a continuous field description: stresses and averaging aspects. Model. Simul. Mater. Sci. Eng. **21**(8), 085,006+ (2013)
10. Wulfinghoff, S., Forest, S., Böhlke, T.: Strain gradient plasticity modeling of the cyclic behavior of laminate microstructures. J. Mech. Phys. Solids **79**, 1–20 (2015)
11. Arsenlis, A., Parks, D.M.: Crystallographic aspects of geometrically-necessary and statistically-stored dislocation density. Acta Mater. **47**(5), 1597–1611 (1999)
12. Hochrainer, T.: Thermodynamically consistent continuum dislocation dynamics. J. Mech. Phys. Solids **88**, 12–22 (2016)
13. Kuhlmann-Wilsdorf, D.: Dislocation cells, redundant dislocations and the leds hypothesis. Scripta Materialia **34**(4), 641–650 (1996)
14. Gurtin, M.E.: A gradient theory of single-crystal viscoplasticity that accounts for geometrically necessary dislocations. J. Mech. Phys. Solids **50**(1), 5–32 (2002)
15. Gurtin, M.E., Anand, L.: A theory of strain-gradient plasticity for isotropic, plastically irrotational materials. Part II: Finite deformations. Int. J. Plast. **21**(12), 2297–2318 (2005)
16. Gurtin, M.E.: A finite-deformation, gradient theory of single-crystal plasticity with free energy dependent on densities of geometrically necessary dislocations. Int. J. Plast. **24**(4), 702–725 (2008)
17. Gurtin, M.E.: A finite-deformation, gradient theory of single-crystal plasticity with free energy dependent on the accumulation of geometrically necessary dislocations. Int. J. Plast. **26**(8), 1073–1096 (2010)
18. Silbermann, C.B., Ihlemann, J.: Geometrically linear continuum theory of dislocations revisited from a thermodynamical perspective. Arch. Appl. Mech. **88**(1–2), 141–173 (2017)
19. Pao, Y.H., Wang, L.S., Chen, K.C.: Principle of Virtual Power for thermomechanics of fluids and solids with dissipation. Int. J. Eng. Sci. **49**(12), 1502–1516 (2011)

Chapter 2
Nonlinear Kinematics of a Continuously Dislocated Crystal

Abstract This chapter presents the basics of geometrically nonlinear kinematics for continuously dislocated crystals. It involves the multiplicative split of the deformation and the slip-system-based decomposition of the velocity gradient. Eventually, appropriate measures for the crystal's strain and geometrically necessary dislocation densities are derived.

2.1 Vectors Describing Cubic Crystal Lattices

Throughout this book, a single crystal with a cubic primitive lattice (Pearson symbol cP) is considered [1, p. 22]. On the one hand, this is done for the sake of simplicity and compactness of the model. On the other hand, the cP-crystal structure enables *plane* deformation for *two* active slip systems (in contrast to cF- or cI-crystals). This way, a minimalistic multi-slip scenario can be studied.

The orientation of the cubic elementary cell is characterized by three mutually perpendicular unit vectors $\underline{\hat{a}}_K$, $K \in \{\mathrm{I}, \mathrm{II}, \mathrm{III}\}$ (cf. Fig. 2.1). These lattice vectors define an orthonormal coordinate system. They are always constant with respect to time and, within a crystallite/single crystal, also with respect to space. To describe anisotropic plastic flow, slip system tensors are introduced, which follow from the knowledge of the lattice vectors. In cP-crystals these are six tensors of the form [1]

$$\underline{\underline{\hat{S}}} = \underline{\hat{s}} \otimes \underline{\widehat{m}} = \underline{\hat{a}}_J \otimes \underline{\hat{a}}_K \quad \text{for all} \quad J \neq K. \tag{2.1}$$

Here, $\underline{\hat{s}}$ represents the slip direction and $\underline{\widehat{m}}$ denotes the slip plane's normal unit vector (NUV). Physically, $\underline{\widehat{m}}$ corresponds to the most densely packed crystal planes and $\underline{\hat{s}}$ follows from the shortest interatomic distances in these planes. Due to the comparatively small number of slip systems, a cP-crystal may not be able to perform all possible plastic distortions. Still, having six (or, in the plane case, two) slip systems represents a significant increase in slip possibilities compared to single-slip conditions provoking extreme plastic anisotropy [2].

C. B. Silbermann et al., *Introduction to Geometrically Nonlinear Continuum Dislocation Theory*, SpringerBriefs in Continuum Mechanics,
https://doi.org/10.1007/978-3-030-63696-8_2

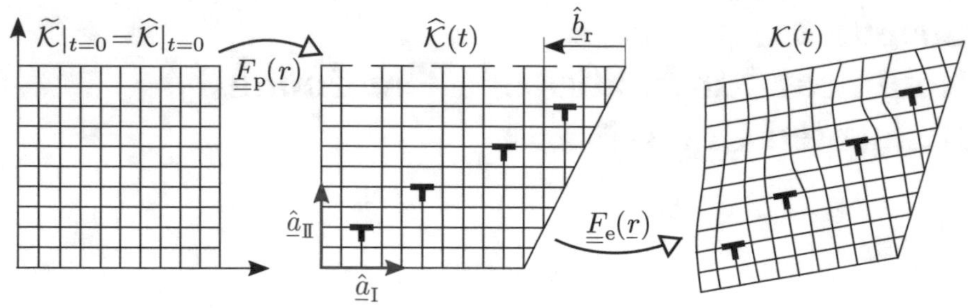

Fig. 2.1 Effect of the fields $\underline{\underline{F}}_{\mathrm{p}}$, $\underline{\underline{F}}_{\mathrm{e}}$ on the crystal's shape and lattice: the resulting Burgers vector $\hat{\underline{b}}_{\mathrm{r}}$ represents the closure failure of the Burgers circuit around $\widehat{\mathcal{K}}(t)$

2.2 Multiplicative Split of the Deformation and Physical Interpretation

From the displacement gradient $\underline{\underline{\beta}} = (\underline{\widetilde{\nabla}} \otimes \underline{u})^{\mathrm{T}}$ immediatly follows the deformation $\underline{\underline{F}} = \underline{\underline{\beta}} + \underline{\underline{I}}$, which can be split *multiplicatively* into elastic and plastic parts:

$$\underline{\underline{F}} = \underline{\underline{F}}_{\mathrm{e}} \cdot \underline{\underline{F}}_{\mathrm{p}} \,. \tag{2.2}$$

In the context of dislocation-based crystal plasticity, a clear physical interpretation can be attributed to this decomposition [3]:

- $\underline{\underline{F}}_{\mathrm{p}}$ comprises all plastic slip processes (due to dislocation motion), whereby the crystal lattice and the lattice vectors do not change.
- $\underline{\underline{F}}_{\mathrm{e}}$ comprises the purely elastic distortion of the crystal lattice, all dislocations being considered "frozen", i.e. the dislocation state remains the same.

The formation of closure failures can (mentally) be attributed to $\underline{\underline{F}}_{\mathrm{p}}$ (cf. Fig. 2.1). From Eq. (2.2) follows that $\underline{\underline{F}}_{\mathrm{e}}$ & $\underline{\underline{F}}_{\mathrm{p}}$ are either both compatible or both incompatible.[1] Thus, both carry the information about those dislocations that are geometrically necessary to make the deformation field $\underline{\underline{F}}$ overall compatible again, so that the cohesion of the crystal in $\mathcal{K}(t)$ is preserved (cf. also Eq. (2.18)).

It is common to associate decomposition (2.2) with the introduction of the new intermediate configuration $\widehat{\mathcal{K}}$, as done in purely phenomenological viscoplasticity (cf. e.g. [4, 5]). Strictly speaking, it is not a *configuration*, it is a *space* [6]: $\underline{\underline{F}}_{\mathrm{e}}$ and $\underline{\underline{F}}_{\mathrm{p}}$ merely transform vectors and do not map from one configuration to another. Only the deformation field $\underline{\underline{F}}$ as gradient of the displacement field is such a mapping. In this sense, $\widehat{\mathcal{K}}$ stands in the context of dislocation-based plasticity for the space or image domain of the lattice that remains unaffected by *plastic* distortion [7, p. 190 f.].

[1] If decomposition (2.2) corresponded to a real (temporal) sequence, the incompatibility of $\underline{\underline{F}}_{\mathrm{e}}$ could be seen as a reaction to that of $\underline{\underline{F}}_{\mathrm{p}}$. In reality, however, there is no such order.

However, the continuum has no lattice *configuration* of its own. This distinction is meaningful, but has no effect on the mathematical structure of the theory. For the sake of clarity, the symbol $\widehat{\mathcal{K}}$ is still used for the lattice space.

2.3 Transformation Rules for Lattice Vectors

Lattice vectors should transform only by the elastic part of the deformation. For this reason all lattice vectors were defined with respect to the lattice space $\widehat{\mathcal{K}}$ [6]. This reflects the fact that a purely plastic distortion does not change the crystal lattice [8]. The natural transformation from $\widehat{\mathcal{K}}$ to the current configuration $\mathcal{K}$ is

$$\underline{a}_K = \underline{\underline{F}}_{\mathrm{e}} \cdot \underline{\hat{a}}_K \ , \ \ \underline{s}_i = \underline{\underline{F}}_{\mathrm{e}} \cdot \underline{\hat{s}}_i \ , \ \ \underline{m}_i = \underline{\underline{F}}_{\mathrm{e}}^{-\mathrm{T}} \cdot \underline{\widehat{m}}_i \ , \ \ \underline{n}\,\mathrm{d}A = I_3\big(\underline{\underline{F}}_{\mathrm{e}}\big)\,\underline{\underline{F}}_{\mathrm{e}}^{-\mathrm{T}} \cdot \underline{\hat{n}}\,\mathrm{d}\widehat{A} \ . \tag{2.3}$$

Consequently, an elastic distortion can change the magnitude and direction of lattice vectors, but the associated slip directions and normal vectors always remain perpendicular to each other, i. e. $\underline{s}_i \cdot \underline{m}_i = 0$. This reflects the fact that dislocations can only move in the crystallographically defined slip planes. Furthermore, the lattice vectors can be transformed into a fictitious reference configuration $\widetilde{\mathcal{K}}$:

$$\underline{\tilde{a}}_K = \underline{\underline{F}}_{\mathrm{p}}^{-1} \cdot \underline{\hat{a}}_K \ , \ \ \underline{\tilde{s}}_i = \underline{\underline{F}}_{\mathrm{p}}^{-1} \cdot \underline{\hat{s}}_i \ , \ \ \underline{\widetilde{m}}_i = \underline{\underline{F}}_{\mathrm{p}}^{\mathrm{T}} \cdot \underline{\widehat{m}}_i \ , \ \ \underline{\tilde{n}}\,\mathrm{d}\widetilde{A} = I_3\Big(\underline{\underline{F}}_{\mathrm{p}}^{-1}\Big)\,\underline{\underline{F}}_{\mathrm{p}}^{\mathrm{T}} \cdot \underline{\hat{n}}\,\mathrm{d}\widehat{A} \ . \tag{2.4}$$

Since the lattice is defined with respect to $\widehat{\mathcal{K}}$ and evolves in the current configuration $\mathcal{K}$, $\widetilde{\mathcal{K}}$ has no physical meaning except in the initial state, when $\widetilde{\mathcal{K}}(t=0) = \widehat{\mathcal{K}}(t=0)$.[2] The introduction of the lattice space proves to be advantageous because the lattice vectors there (and only there) are constant. This property is lost as soon as they are mapped to $\widetilde{\mathcal{K}}$ or $\mathcal{K}$. For the transformation of infinitesimal material line and surface elements from $\widetilde{\mathcal{K}}$ onto $\mathcal{K}$ the well-known relations result:

$$\underline{a}_K = \underline{\underline{F}} \cdot \underline{\tilde{a}}_K \ , \ \ \underline{s}_i = \underline{\underline{F}} \cdot \underline{\tilde{s}}_i \ , \ \ \underline{m}_i = \underline{\underline{F}}^{-\mathrm{T}} \cdot \underline{\widetilde{m}}_i \ , \ \ \underline{n}\,\mathrm{d}A = I_3\big(\underline{\underline{F}}\big)\,\underline{\underline{F}}^{-\mathrm{T}} \cdot \underline{\tilde{n}}\,\mathrm{d}\widetilde{A} \ . \tag{2.5}$$

Note that $\underline{n}$, $\underline{\hat{n}}$, $\underline{\tilde{n}}$ are all normal *unit* vectors since they transform according to Nanson-like relations. The transformed slip plane normals $\underline{m}_i$, $\underline{\widetilde{m}}_i$ are no longer unit vectors, neither are the vectors $\underline{s}_i$, $\underline{\tilde{s}}_i$ and $\underline{a}_K$, $\underline{\tilde{a}}_K$.

---

[2] Reference configuration and initial configuration are in general *not* identical [9, p. 20].

2.4 Additive Split of the Velocity Gradient

The multiplicative decomposition (2.2) leads to an *additive* decomposition of the velocity gradient.[3] Using the abbreviations $\underline{\underline{L}}_e = \dot{\underline{\underline{F}}}_e \cdot \underline{\underline{F}}_e^{-1}$ and $\hat{\underline{\underline{L}}}_p = \dot{\underline{\underline{F}}}_p \cdot \underline{\underline{F}}_p^{-1}$ it reads

$$\underline{\underline{L}} = (\underline{\nabla} \otimes \underline{v})^{\mathrm{T}} = \dot{\underline{\underline{F}}} \cdot \underline{\underline{F}} = \underline{\underline{L}}_e + \underline{\underline{F}}_e \cdot \hat{\underline{\underline{L}}}_p \cdot \underline{\underline{F}}_e^{-1} . \tag{2.6}$$

The essential assumption of crystal plasticity is to represent $\hat{\underline{\underline{L}}}_p$ by a sum of different slip system contributions (simple shear) with respect to $\widehat{\mathcal{K}}$, i. e.,

$$\hat{\underline{\underline{L}}}_p = \sum_i \nu_i \, \hat{\underline{s}}_i \otimes \hat{\underline{m}}_i = \sum_i \nu_i \, \hat{\underline{\underline{S}}}_i , \tag{2.7}$$

where ν_i denotes the slip rate in the ith slip system.[4] Inserting this into Eq. (2.6) and considering transformation rule (2.3) it follows:

$$\underline{\underline{L}} = \underline{\underline{L}}_e + \underline{\underline{F}}_e \cdot \sum_i \nu_i \, \hat{\underline{s}}_i \otimes \hat{\underline{m}}_i \cdot \underline{\underline{F}}_e^{-1} = \underline{\underline{L}}_e + \sum_i \nu_i \, \underline{s}_i \otimes \underline{m}_i . \tag{2.8}$$

This shows that the slip systems are subject to elastic distortion, i. e. *not* constant as in the geometrically linear theory [2]. In general, the lattice vectors are elastically distorted and rotated (cf. Fig. 2.1 on the right). Decomposition of Eq. (2.8) into a symmetrical and a skew-symmetrical part results in an additive splitting into a distortion velocity $\underline{\underline{D}}$ and a rotation velocity (spin) $\underline{\underline{\Omega}}$:

$$\underline{\underline{D}} = \mathrm{sym}(\underline{\underline{L}}) = \mathrm{sym}(\underline{\underline{L}}_e) + \tfrac{1}{2} \sum_i \nu_i \left(\underline{s}_i \otimes \underline{m}_i + \underline{m}_i \otimes \underline{s}_i \right) = \underline{\underline{D}}_e + \underline{\underline{D}}_p , \tag{2.9}$$

$$\underline{\underline{\Omega}} = \mathrm{skw}(\underline{\underline{L}}) = \mathrm{skw}(\underline{\underline{L}}_e) + \tfrac{1}{2} \sum_i \nu_i \left(\underline{s}_i \otimes \underline{m}_i - \underline{m}_i \otimes \underline{s}_i \right) = \underline{\underline{\Omega}}_e + \underline{\underline{\Omega}}_p . \tag{2.10}$$

For purely elastic deformation, the lattice spin $\underline{\underline{\Omega}}_e$ equals the total spin $\underline{\underline{\Omega}}$. Through plastic slip processes the body shape changes, which leads to the plastic spin $\underline{\underline{\Omega}}_p$. As in geometrically linear theory, plastic distortion influences the elastic distortion of the crystal lattice. New, however, is the nonlinear feedback of the elastic lattice distortion to the plastic distortion by Relation (2.8).

[3] However, note that $\hat{\underline{\underline{L}}}_p$ is transformed again *multiplicatively* by $\underline{\underline{F}}_e$ such that this is no pure additive decomposition.

[4] The index variables $i, j \in \{1, 2, \dots\}$ mark the slip systems. A summation over slip systems is always indicated by a $\sum$ sign. For i, j there is *no* summation convention.

Remark 2.1 The distortion of slip systems also affects the true slip rates, which follow from scaling $\underline{s}_i \otimes \underline{m}_i$ to unit vectors with

$$\|\underline{s}_i \otimes \underline{m}_i\| = |\underline{s}_i||\underline{m}_i| = \sqrt{\left(\hat{\underline{s}}_i \cdot \hat{\underline{\underline{C}}}_{\mathrm{e}} \cdot \hat{\underline{s}}_i\right)\left(\widehat{\underline{m}}_i \cdot \hat{\underline{\underline{C}}}_{\mathrm{e}}^{-1} \cdot \widehat{\underline{m}}_i\right)} =: R_i(\underline{\underline{F}}_{\mathrm{e}}) \quad (2.11)$$

as $R_i\,\nu_i$. Since the elastic distortion can be assumed small, only ν_i will be used in the following constitutive relations—for simplicity's sake. An enrichment of the nonlinear interactions by $\nu_i \rightarrow R_i(\underline{\underline{F}}_{\mathrm{e}})\,\nu_i$ seems interesting for future research.

The definition $\hat{\underline{\underline{L}}}_{\mathrm{p}} = \dot{\underline{\underline{F}}}_{\mathrm{p}} \cdot \underline{\underline{F}}_{\mathrm{p}}^{-1}$ may be regarded as the ordinary first-order differential equation $\dot{\underline{\underline{F}}}_{\mathrm{p}} = \hat{\underline{\underline{L}}}_{\mathrm{p}} \cdot \underline{\underline{F}}_{\mathrm{p}}$. Inserting the crystal plasticity form (2.7), the evolution of $\underline{\underline{F}}_{\mathrm{p}}$ is obtained as a result of individual slip processes (cf. Sect. 6.3):

$$\dot{\underline{\underline{F}}}_{\mathrm{p}} = \sum_i \nu_i\, \hat{\underline{s}}_i \otimes \widehat{\underline{m}}_i \cdot \underline{\underline{F}}_{\mathrm{p}} \;\rightarrow\; \underline{\underline{F}}_{\mathrm{p}}(t{+}\Delta t) = \exp\left(\!\!\left(\sum_i \nu_i(t{+}\Delta t)\Delta t\; \hat{\underline{s}}_i \otimes \widehat{\underline{m}}_i\right)\!\!\right) \cdot \underline{\underline{F}}_{\mathrm{p}}(t)\,. \quad (2.12)$$

Taking into account the orthogonality between slip plane and direction, decomposition (2.7) and accordingly Formula (2.8) has the important consequence

$$I_1(\hat{\underline{\underline{L}}}_{\mathrm{p}}) = 0 \quad \text{and} \quad I_1(\underline{\underline{L}}) = I_1(\underline{\underline{L}}_{\mathrm{e}})\,. \quad (2.13)$$

Under the initial condition $I_3(\underline{\underline{F}}_{\mathrm{p}}(t=0)) = 1$, it follows that $I_3(\underline{\underline{F}}_{\mathrm{p}}) = 1$ and $\underline{\underline{F}}_{\mathrm{p}}$ is a unimodular tensor. From the physical point of view: the plastic flow is isochoric.[5] Thus, the volumetric strain can be taken from $\underline{\underline{F}}_{\mathrm{e}}$.

2.5 Measures for Elastic Lattice Strain and -Rotation

From tensor $\underline{\underline{F}}_{\mathrm{e}}$ some measures for elastic stretch and strain can be calculated as

$$\underline{\underline{b}}_{\mathrm{e}} = \underline{\underline{F}}_{\mathrm{e}} \cdot \underline{\underline{F}}_{\mathrm{e}}^{\mathrm{T}}, \quad \hat{\underline{\underline{C}}}_{\mathrm{e}} = \underline{\underline{F}}_{\mathrm{e}}^{\mathrm{T}} \cdot \underline{\underline{F}}_{\mathrm{e}}, \quad \hat{\underline{\underline{E}}}_{\mathrm{e}} = \tfrac{1}{2}(\hat{\underline{\underline{C}}}_{\mathrm{e}} - \underline{\underline{I}}), \quad j = \frac{\mathrm{d}V}{\mathrm{d}\widetilde{V}} = I_3(\underline{\underline{F}}) = I_3(\underline{\underline{F}}_{\mathrm{e}})\,. \quad (2.14)$$

[5]This meets the expectation as long as only wandering of dislocations in the slip planes is taken into account, but not the possible climbing of edge dislocations under thermal activation.

The tensors $\hat{\underline{\underline{C}}}_{\mathrm{e}}$ and $\hat{\underline{\underline{E}}}_{\mathrm{e}}$ measure the elastic lattice stretching and distortion respectively and are elements of the lattice space $\widehat{\mathcal{K}}$. For a cubic crystal, the elastic lattice strains ε_K and angular changes $\varepsilon_{J,K}$ result from the current lattice vectors:

$$\varepsilon_K = \underline{\hat{a}}_K \cdot \hat{\underline{\underline{E}}}_{\mathrm{e}} \cdot \underline{\hat{a}}_K = \tfrac{1}{2}\left\{|\underline{a}_K|^2 - 1\right\}, \quad \varepsilon_{J,K} = \underline{\hat{a}}_J \cdot \hat{\underline{\underline{E}}}_{\mathrm{e}} \cdot \underline{\hat{a}}_K = \tfrac{1}{2}\left\{\underline{a}_J \cdot \underline{a}_K\right\} \text{ for } J \neq K .$$

In the case of plane deformation (e. g. in the plane perpendicular to $\underline{\hat{a}}_{\mathrm{III}}$) this means:

$$\varepsilon_{\mathrm{I}} = \tfrac{1}{2}\left\{|\underline{a}_{\mathrm{I}}|^2 - 1\right\} \quad , \quad \varepsilon_{\mathrm{II}} = \tfrac{1}{2}\left\{|\underline{a}_{\mathrm{II}}|^2 - 1\right\} \quad , \quad \varepsilon_{\mathrm{I,II}} = \tfrac{1}{2}\left\{\underline{a}_{\mathrm{I}} \cdot \underline{a}_{\mathrm{II}}\right\} . \qquad (2.15)$$

Applying the polar decomposition [9, p. 26] separately to both parts of $\underline{\underline{F}}$, elastic and plastic contributions of the rotation can be obtained:

$$\underline{\underline{F}} = \underline{\underline{R}} \cdot \widetilde{\underline{\underline{U}}} = \underline{\underline{F}}_{\mathrm{e}} \cdot \underline{\underline{F}}_{\mathrm{p}} = \left(\underline{\underline{R}}_{\mathrm{e}} \cdot \underline{\underline{U}}_{\mathrm{e}}\right) \cdot \left(\underline{\underline{R}}_{\mathrm{p}} \cdot \widetilde{\underline{\underline{U}}}_{\mathrm{p}}\right) = \hat{\underline{\underline{V}}}_{\mathrm{e}} \cdot \left(\underline{\underline{R}}_{\mathrm{e}} \cdot \underline{\underline{R}}_{\mathrm{p}}\right) \cdot \widetilde{\underline{\underline{U}}}_{\mathrm{p}} . \qquad (2.16)$$

Transformation (2.3) shows that the "elastic" part $\underline{\underline{R}}_{\mathrm{e}}$ describes the rotation of the crystal lattice (inner rotation). The other part measures the rotation of the outer contour/shape of the body due to plastic slip (outer rotation). It is remarkable that the total rigid body rotation $\underline{\underline{R}}$ is not equal to the product $\underline{\underline{R}}_{\mathrm{e}} \cdot \underline{\underline{R}}_{\mathrm{p}}$. A clearer interpretation is given by the additive decomposition of the rotational velocities in Eq. (2.10).

2.6 Measures for Geometrically Necessary Dislocations

The definition of the resulting Burgers vector is similar to the geometrically linear theory [2]. A closed-loop path integral around a crystal (cut) surface $\widehat{A}$ with the contour $\widehat{C} = \partial\widehat{A}$ in the lattice space $\widehat{\mathcal{K}}$ (Burgers circuit) results in a closure failure:

$$\underline{\hat{b}}_{\mathrm{r}} = \oint_{\widehat{C}} \mathrm{d}\underline{\hat{r}} = \begin{cases} \oint_{\widehat{C}} \underline{\underline{F}}_{\mathrm{p}} \cdot \mathrm{d}\underline{\tilde{r}} = \int_{\widetilde{A}} \left(-\underline{\underline{F}}_{\mathrm{p}} \times \widetilde{\underline{\nabla}}\right) \cdot \underline{\tilde{n}}\, \mathrm{d}\widetilde{A} = \int_{\widehat{A}} \mathrm{Curl}\left(\underline{\underline{F}}_{\mathrm{p}}\right) \cdot I_3\left(\underline{\underline{F}}_{\mathrm{p}}^{-1}\right) \underline{\underline{F}}_{\mathrm{p}}^{\mathrm{T}} \cdot \underline{\hat{n}}\, \mathrm{d}\widehat{A} \\ \oint_{\widehat{C}} \underline{\underline{F}}_{\mathrm{e}}^{-1} \cdot \mathrm{d}\underline{r} = \int_{A} \left(-\underline{\underline{F}}_{\mathrm{e}}^{-1} \times \underline{\nabla}\right) \cdot \underline{n}\, \mathrm{d}A = \int_{\widehat{A}} \mathrm{curl}\left(\underline{\underline{F}}_{\mathrm{e}}^{-1}\right) \cdot I_3\left(\underline{\underline{F}}_{\mathrm{e}}\right) \underline{\underline{F}}_{\mathrm{e}}^{-\mathrm{T}} \cdot \underline{\hat{n}}\, \mathrm{d}\widehat{A} \end{cases} .$$

The realization of this Burgers circuit with respect to the incompatible lattice space is also reasonable because there are ideal lattice vectors which are neither distorted nor rotated. The transformation of the contour integral to the *compatible* reference or current configuration is necessary to apply Stokes' integral theorem. The subsequent inverse transformation into the lattice space allows the comparability of both integrands:

$$\underline{\hat{b}}_{\mathrm{r}} = \oint_{\widehat{C}} \mathrm{d}\underline{\hat{r}} = \int_{\hat{A}} \hat{\underline{\underline{\alpha}}} \cdot \underline{\hat{n}}\, \mathrm{d}\hat{A} \quad \text{and} \quad \mathrm{d}\underline{\hat{b}}_{\mathrm{r}} = \hat{\underline{\underline{\alpha}}} \cdot \underline{\hat{n}}\, \mathrm{d}\hat{A} . \qquad (2.17)$$

From this follows the second-order tensor $\underline{\underline{\hat{\alpha}}}$, which obviously corresponds to a line density (dimension length/area) [10]. Exploiting Eq. $(2.14)_4$ it can be calculated as

$$\underline{\underline{\hat{\alpha}}} := \operatorname{Curl}\left(\underline{\underline{F}}_{\mathrm{p}}\right) \cdot \underline{\underline{F}}_{\mathrm{p}}^{\mathrm{T}} = j \operatorname{curl}\left(\underline{\underline{F}}_{\mathrm{e}}^{-1}\right) \cdot \underline{\underline{F}}_{\mathrm{e}}^{-\mathrm{T}} . \tag{2.18}$$

As in geometrically linear theory, the dislocation density tensor can be calculated from both elastic and plastic distortion, owed to the compatibility of the deformation field. Furthermore, $\underline{\underline{\hat{\alpha}}}$ was defined with respect to the lattice space. The transformation into the current configuration is:

$$\underline{\underline{\alpha}} = \underline{\underline{F}}_{\mathrm{e}} \cdot \underline{\underline{\hat{\alpha}}} \cdot \underline{\underline{F}}_{\mathrm{e}}^{\mathrm{T}} = j\, \underline{\underline{F}}_{\mathrm{e}} \cdot \operatorname{curl}\left(\underline{\underline{F}}_{\mathrm{e}}^{-1}\right) = j\, \underline{\underline{F}}_{\mathrm{e}} \cdot \left(-\underline{\underline{F}}_{\mathrm{e}}^{-1} \times \underline{\nabla}\right) . \tag{2.19}$$

The evolution of the geometrically necessary dislocation state follows from the rate

$$\dot{\underline{\underline{\hat{\alpha}}}} \overset{(2.18)}{=} \left(\left(-\underline{\underline{F}}_{\mathrm{p}} \times \underline{\widetilde{\nabla}}\right) \cdot \underline{\underline{F}}_{\mathrm{p}}^{\mathrm{T}}\right)^{\cdot} = \left(-\underline{\underline{F}}_{\mathrm{p}} \times \underline{\widetilde{\nabla}}\right) \cdot \dot{\underline{\underline{F}}}_{\mathrm{p}}^{\mathrm{T}} + \left(-\underline{\underline{F}}_{\mathrm{p}} \times \underline{\widetilde{\nabla}}\right)^{\cdot} \cdot \underline{\underline{F}}_{\mathrm{p}}^{\mathrm{T}} . \tag{2.20}$$

Note that the order of differentiation with respect to referential coordinates and time is interchangeable. Thus, with $(\underline{\widetilde{\nabla}})^{\cdot} = \underline{0}$ and $\dot{\underline{\underline{F}}}_{\mathrm{p}} \times \underline{\widetilde{\nabla}} = (\underline{\underline{\hat{L}}}_{\mathrm{p}} \cdot \underline{\underline{F}}_{\mathrm{p}}) \times \underline{\widetilde{\nabla}}$ it follows

$$\dot{\underline{\underline{\hat{\alpha}}}} = \underline{\underline{\hat{\alpha}}} \cdot \underline{\underline{\hat{L}}}_{\mathrm{p}}^{\mathrm{T}} + \underline{\underline{\hat{L}}}_{\mathrm{p}} \cdot \underline{\underline{\hat{\alpha}}} + (\widetilde{\nabla}_a \underline{\underline{\hat{L}}}_{\mathrm{p}}) \cdot (-\underline{\underline{F}}_{\mathrm{p}} \times \underline{e}_a) \cdot \underline{\underline{F}}_{\mathrm{p}}^{\mathrm{T}} . \tag{2.21}$$

For further simplification, the slip system approach (2.7) is used again, resulting in

$$\dot{\underline{\underline{\hat{\alpha}}}} = \sum_i \nu_i \left(\underline{\underline{\hat{\alpha}}} \cdot \underline{\underline{\hat{S}}}_i^{\mathrm{T}} + \underline{\underline{\hat{S}}}_i \cdot \underline{\underline{\hat{\alpha}}}\right) + \sum_i \underline{\underline{\hat{S}}}_i \cdot (-\underline{\underline{F}}_{\mathrm{p}} \times \underline{\widetilde{\nabla}} \nu_i) \cdot \underline{\underline{F}}_{\mathrm{p}}^{\mathrm{T}} . \tag{2.22}$$

By aid of algebraic Relation (A.4) and $I_3\left(\underline{\underline{F}}_{\mathrm{p}}\right) = 1$ as well as $\underline{\widetilde{\nabla}} = \underline{\underline{F}}^{\mathrm{T}} \cdot \underline{\nabla}$ it follows

$$\underline{\underline{\hat{S}}}_i \cdot \left(-\underline{\underline{F}}_{\mathrm{p}} \times \underline{\widetilde{\nabla}} \nu_i\right) \cdot \underline{\underline{F}}_{\mathrm{p}}^{\mathrm{T}} = \underline{\underline{\hat{S}}}_i \cdot \underline{\underline{\underline{\epsilon}}} \cdot (\underline{\underline{F}}_{\mathrm{p}}^{-\mathrm{T}} \cdot \underline{\widetilde{\nabla}} \nu_i) = -\underline{\underline{\hat{S}}}_i \times (\underline{\underline{F}}_{\mathrm{e}}^{\mathrm{T}} \cdot \underline{\nabla} \nu_i) = \underline{\hat{s}}_i \otimes \underline{\widehat{\nabla}} \nu_i \times \underline{\widehat{m}}_i , \tag{2.23}$$

where $\underline{\widehat{\nabla}} \nu_i := \underline{\underline{F}}_{\mathrm{e}}^{\mathrm{T}} \cdot \underline{\nabla} \nu_i$ denotes the slip rate gradient transformed into the lattice space. At this point the correspondence to the geometrically linear theory becomes clear, where on *position* level a completely analogous relation holds [2]. In contrast, term (2.23) is on *rate* level and the temporal change of $\underline{\underline{\hat{\alpha}}}$ is not only determined by it, but reads

$$\dot{\underline{\underline{\hat{\alpha}}}} = \sum_i \underline{\hat{s}}_i \otimes \underline{\widehat{\nabla}} \nu_i \times \underline{\widehat{m}}_i + \sum_i \nu_i \left(\underline{\underline{\hat{\alpha}}} \cdot \underline{\underline{\hat{S}}}_i^{\mathrm{T}} + \underline{\underline{\hat{S}}}_i \cdot \underline{\underline{\hat{\alpha}}}\right) . \tag{2.24}$$

This important difference can be considered as an effect of the deforming slip systems: Both slip rate gradients cause a change of the dislocation state as well as gradients of the lattice distortion. Accordingly, $\underline{\underline{\hat{\alpha}}}$ and its invariants contain significantly more information than $\underline{\nabla} v_i$ alone. The dislocation density tensor contains information about the local closure failure at any crystal plane (cf. e. g. [11]). To assign the closure failure to a particular type of dislocation, the Burgers circuit must be performed in a crystal plane perpendicular to the tangent lines. The resulting dislocation density *vector* (cf. e. g. [11]) can then be decomposed into a normal portion with screw character and two tangential portions with edge character (cf. e. g. [12]). In the present work, a 2D analysis in the xy-plane is conducted, leaving only one cut plane with $\hat{\underline{n}} = \underline{e}_z$. This further implies *straight edge* dislocations for whose elementary Burgers vector and tangent vector it applies

$$\hat{\underline{b}}_i = b\,\hat{\underline{s}}_i \quad , \quad \hat{\underline{t}}_i = \hat{\underline{s}}_i \times \hat{\underline{m}}_i \; . \tag{2.25}$$

This also provides the necessary sign convention to always measure the Burgers vector in the same direction of rotation: It is agreed that the Burgers circuit is mathematically positive (right hand rule) in relation to $\hat{\underline{t}}_i$. The two-dimensional consideration of a cP-crystal leaves out of the six slip systems (2.1) only two with $\hat{\underline{t}}_1 = -\hat{\underline{a}}_{\mathrm{III}} = -\underline{e}_z$ and $\hat{\underline{t}}_2 = \hat{\underline{a}}_{\mathrm{III}} = \underline{e}_z$. The corresponding dislocation densities are:

$$\rho_1 = \frac{1}{b}\,\hat{\underline{s}}_1 \cdot \hat{\underline{\underline{\alpha}}} \cdot \hat{\underline{t}}_1 \qquad = -\frac{1}{b}\,\hat{\underline{a}}_{\mathrm{II}} \cdot \left(\hat{\underline{\underline{\alpha}}} \cdot \underline{e}_z\right) = -\frac{1}{b}\,\hat{\underline{a}}_{\mathrm{II}} \cdot \hat{\underline{\alpha}}_z \tag{2.26a}$$

$$\rho_2 = \frac{1}{b}\,\hat{\underline{s}}_2 \cdot \hat{\underline{\underline{\alpha}}} \cdot \hat{\underline{t}}_2 \qquad = \frac{1}{b}\,\hat{\underline{a}}_{\mathrm{I}} \cdot \left(\hat{\underline{\underline{\alpha}}} \cdot \underline{e}_z\right) = \frac{1}{b}\,\hat{\underline{a}}_{\mathrm{I}} \cdot \hat{\underline{\alpha}}_z . \tag{2.26b}$$

These scalar, *signed* quantities represent invariants of the dislocation density tensor. In a 2D view, obviously only the information of a dislocation density vector $\hat{\underline{\alpha}}_z$ with respect to the cut face normal $\hat{\underline{n}} = \underline{e}_z$ is needed. Due to the orthogonality of the slip systems in the cP-crystal, the sum of squared dislocation densities yields

$$b^2\left(\rho_1^2 + \rho_2^2\right) = \|\hat{\underline{\underline{\alpha}}}\|^2 \; . \tag{2.27}$$

In order to evaluate the plausibility of the introduced dislocation densities all quantities involved are transformed onto $\mathcal{K}$ taking into account rule (2.3) and (2.19):

$$b\rho_i = \hat{\underline{s}}_i \cdot \left(\underline{\underline{F}}_{\mathrm{e}}^{\mathrm{T}} \cdot \underline{\underline{F}}_{\mathrm{e}}^{-\mathrm{T}}\right) \cdot \left(\underline{\underline{F}}_{\mathrm{e}}^{-1} \cdot \underline{\underline{F}}_{\mathrm{e}}\right) \cdot \hat{\underline{\underline{\alpha}}} \cdot \left(\underline{\underline{F}}_{\mathrm{e}}^{\mathrm{T}} \cdot \underline{\underline{F}}_{\mathrm{e}}^{-\mathrm{T}}\right) \cdot \hat{\underline{t}}_i = \underline{s}_i \cdot \left(\underline{\underline{b}}_{\mathrm{e}}^{-1} \cdot \underline{\underline{\alpha}}\right) \cdot \underline{t}_i \; .$$

Obviously, these invariants contain not only information about the dislocation density, but also about the distortion within the cut plane with normal $\underline{t}_i$. This merging originates from the transformation rules (2.3) and is explained by the fact that the vectors $\underline{s}$ and $\underline{t}$ do not represent unit vectors anymore in $\mathcal{K}$ (cf. Remark 2.1).

References

1. Kleber, W., Bautsch, H.J., Bohm, J.: Einführung in die Kristallographie. Oldenbourg Verlag, München (1998)
2. Silbermann, C.B., Ihlemann, J.: Geometrically linear continuum theory of dislocations revisited from a thermodynamical perspective. Arch. Appl. Mech. **88**(1–2), 141–173 (2017)
3. Le, K.C., Günther, C.: Nonlinear continuum dislocation theory revisited. Int. J. Plast. **53**, 164–178 (2014)
4. Shutov, A.V., Kuprin, C., Ihlemann, J., Wagner, M.F.X., Silbermann, C.: Experimentelle Untersuchung und numerische Simulation des inkrementellen Umformverhaltens von Stahl 42CrMo4. Materialwiss. Werkstofftech. **41**(9), 765–775 (2010)
5. Silbermann, C.B., Shutov, A.V., Ihlemann, J.: On operator split technique for the time integration within finite strain viscoplasticity in explicit FEM. PAMM **14**(1), 355–356 (2014)
6. Gurtin, M.E., Reddy, B.D.: Some issues associated with the intermediate space in single-crystal plasticity. J. Mech. Phys. Solids **95**, 230–238 (2016)
7. Krawietz, A.: Materialtheorie : Mathematische Beschreibung des phänomenologischen thermomechanischen Verhaltens. Springer, New York, Tokyo (1986)
8. Kröner, E.: Kontinuumstheorie der Versetzungen und Eigenspannungen. Springer, Berlin (1958)
9. Wriggers, P.: Nonlinear Finite Element Methods. Springer, Berlin (2008)
10. Gurtin, M.E.: A gradient theory of single-crystal viscoplasticity that accounts for geometrically necessary dislocations. J. Mech. Phys. Solids **50**(1), 5–32 (2002)
11. Silbermann, C.B., Ihlemann, J.: Analogies between continuum dislocation theory, continuum mechanics and fluid mechanics. IOP Conf. Ser. Mater. Sci. Eng. **181**, 012,037+ (2017)
12. Silbermann, C.B., Ihlemann, J.: Kinematic assumptions and their consequences on the structure of field equations in continuum dislocation theory. IOP Conf. Ser. Mater. Sci. Eng. **118**, 012, 034+ (2016)

Chapter 3
Crystal Kinetics and -Thermodynamics

Abstract This chapter presents formulations for the free energy attributed to elastic strains and geometrically necessary dislocations of a cubic (primitive) crystal. Proposing a consistent thermodynamical framework, field equations are derived, starting from the Clausius-Planck inequality. Thereby, energy dissipation due to dislocation motion is introduced in a thermodynamically consistent way. Finally, different flow rules for the slip systems are presented and discussed with regard to the thermodynamic consequences.

3.1 Additive Split of the Crystal's Free Energy

Let the inner mechanical state be characterized by the elastic strain tensor and the dislocation density tensor, the inner thermal state by the absolute temperature θ. Then, the specific free energy of the crystal depends only on these thermodynamic state variables. An additive decomposition is assumed [1]:

$$\psi = \psi(\underline{\underline{\hat{E}}}_{\mathrm{e}}, \underline{\underline{\hat{\alpha}}}, \theta) = \psi_{\mathrm{e}}(\underline{\underline{\hat{E}}}_{\mathrm{e}}) + \psi_{\mathrm{p}}(\underline{\underline{\hat{\alpha}}}) + \psi_{\mathrm{t}}(\theta) \,. \tag{3.1}$$

The exclusive use of the arguments given in the form (3.1) makes the material behavior become objective (i. e. 'reference frame'-invariant) and observer-independent in the sense of [2].[1] The integration of the specific free energy over the mass[2] yields the total free energy

$$\mathcal{F} = \int\limits_M \psi \,\mathrm{d}m = \int\limits_V \varrho\psi \,\mathrm{d}V = \int\limits_V \phi \,\mathrm{d}V \tag{3.2}$$

[1] All quantities from $\widehat{\mathcal{K}}$ exhibit a 'reference frame'-invariance and observer-independence [3, 4].

[2] As mass constancy can be assumed here, the actual mass is equal to the reference mass.

C. B. Silbermann et al., *Introduction to Geometrically Nonlinear Continuum Dislocation Theory*, SpringerBriefs in Continuum Mechanics,
https://doi.org/10.1007/978-3-030-63696-8_3

as integral of the free energy density $\phi = \varrho\psi$ with ϱ denoting the current mass density. Furthermore, it is practical to introduce a free energy density $\hat{\phi}$ related to the *constant* reference volume. Due to the plastic incompressibility (2.14)$_4$ $\hat{\phi} = \hat{\varrho}\psi = \tilde{\varrho}\psi = \tilde{\phi}$ holds and it follows:

$$\mathcal{F} = \int\limits_V \phi \, \mathrm{d}V = \int\limits_V \left(\frac{\varrho}{\tilde{\varrho}}\right) \tilde{\phi} \, \mathrm{d}V = \int\limits_V \frac{1}{j} \hat{\phi} \, \mathrm{d}V \, . \tag{3.3}$$

3.2 Strain Energy for Cubic-Anisotropic Crystal Elasticity

The elastic behavior of crystals is significantly influenced by the lattice structure and its symmetry. The specific elastic strain energy ψ_e therefore depends not only on the lattice strain but also on the lattice structure. In order to obtain a quadratic strain energy, nine independent invariants can be formed from the products of the elastic lattice strains (2.15). These invariants can be represented by the double contractions of the elastic strain tensor with the structural tensors $\underline{\underline{\hat{A}}}_K = \underline{\hat{a}}_K \otimes \underline{\hat{a}}_K$:

$$(\underline{\underline{\hat{A}}}_K \cdot\cdot\, \underline{\underline{\hat{E}}}_\mathrm{e})^2 = \varepsilon_K^2, \quad (\underline{\underline{\hat{A}}}_J \cdot\cdot\, \underline{\underline{\hat{E}}}_\mathrm{e})(\underline{\underline{\hat{A}}}_K \cdot\cdot\, \underline{\underline{\hat{E}}}_\mathrm{e}) = \varepsilon_J \varepsilon_K, \quad (\underline{\underline{\hat{A}}}_J \cdot \underline{\underline{\hat{E}}}_\mathrm{e}) \cdot\cdot\, (\underline{\underline{\hat{A}}}_K \cdot \underline{\underline{\hat{E}}}_\mathrm{e}) = \varepsilon_{J,K}^2 \, .$$

With this, a quadratic strain energy density suitable for cubic anisotropic crystal lattices can be formulated:[3]

$$\hat{\phi}_\mathrm{e} = \frac{\varkappa}{2} \left(\varepsilon_\mathrm{I}^2 + \varepsilon_\mathrm{II}^2 + \varepsilon_\mathrm{III}^2\right) + \lambda \left(\varepsilon_\mathrm{I}\varepsilon_\mathrm{II} + \varepsilon_\mathrm{II}\varepsilon_\mathrm{III} + \varepsilon_\mathrm{III}\varepsilon_\mathrm{I}\right) + \mu \left(\varepsilon_\mathrm{I,II}^2 + \varepsilon_\mathrm{II,III}^2 + \varepsilon_\mathrm{III,I}^2\right) . \tag{3.4}$$

From this the *constant, anisotropic* elasticity tensor $\underline{\underline{\underline{\underline{\hat{K}}}}} = \dfrac{\partial^2 \hat{\phi}_\mathrm{e}}{\partial \underline{\underline{\hat{E}}}_\mathrm{e} \, \partial \underline{\underline{\hat{E}}}_\mathrm{e}}$ can be derived:

$$\begin{aligned}
\underline{\underline{\underline{\underline{\hat{K}}}}} = \varkappa &\left\{\underline{\underline{\hat{A}}}_\mathrm{I} \otimes \underline{\underline{\hat{A}}}_\mathrm{I} + \underline{\underline{\hat{A}}}_\mathrm{II} \otimes \underline{\underline{\hat{A}}}_\mathrm{II} + \underline{\underline{\hat{A}}}_\mathrm{III} \otimes \underline{\underline{\hat{A}}}_\mathrm{III}\right\} \\
+ \lambda &\left\{\underline{\underline{\hat{A}}}_\mathrm{I} \otimes \underline{\underline{\hat{A}}}_\mathrm{II} + \underline{\underline{\hat{A}}}_\mathrm{II} \otimes \underline{\underline{\hat{A}}}_\mathrm{I} + \underline{\underline{\hat{A}}}_\mathrm{II} \otimes \underline{\underline{\hat{A}}}_\mathrm{III} + \underline{\underline{\hat{A}}}_\mathrm{III} \otimes \underline{\underline{\hat{A}}}_\mathrm{II} + \underline{\underline{\hat{A}}}_\mathrm{III} \otimes \underline{\underline{\hat{A}}}_\mathrm{I} + \underline{\underline{\hat{A}}}_\mathrm{I} \otimes \underline{\underline{\hat{A}}}_\mathrm{III}\right\} \\
+ \mu &\left\{\underline{\underline{\hat{A}}}_\mathrm{I} \otimes \underline{\underline{\hat{A}}}_\mathrm{II} + \underline{\underline{\hat{A}}}_\mathrm{II} \otimes \underline{\underline{\hat{A}}}_\mathrm{I} + \underline{\underline{\hat{A}}}_\mathrm{II} \otimes \underline{\underline{\hat{A}}}_\mathrm{III} + \underline{\underline{\hat{A}}}_\mathrm{III} \otimes \underline{\underline{\hat{A}}}_\mathrm{II} + \underline{\underline{\hat{A}}}_\mathrm{III} \otimes \underline{\underline{\hat{A}}}_\mathrm{I} + \underline{\underline{\hat{A}}}_\mathrm{I} \otimes \underline{\underline{\hat{A}}}_\mathrm{III}\right\}^{\mathrm{S}_{24}} .
\end{aligned} \tag{3.5}$$

The S_{24}-operator defined by Formula (A.5) ensures appropriate symmetry properties of the fourth-order elasticity tensor. Using these symmetries, its coefficients can be represented compactly in a symmetric 6×6–Matrix [2, p. 30 f.]:

[3] The formulation with the structural tensors $\underline{\underline{\hat{A}}}_K$ makes ψ_e an *isotropic* tensor function. Thereby, Relation (3.4) combined with (3.29) fulfills the axiom of material objectivity [5, p. 459].

$$\underline{\underline{\underline{\underline{\widehat{K}}}}} = \widehat{K}_{\alpha\beta}\, \underline{\underline{j}}_\alpha \otimes \underline{\underline{j}}_\beta \quad \text{with} \quad [\widehat{K}_{\alpha\beta}] = \begin{bmatrix} \varkappa & \lambda & \lambda & 0 & 0 & 0 \\ \lambda & \varkappa & \lambda & 0 & 0 & 0 \\ \lambda & \lambda & \varkappa & 0 & 0 & 0 \\ 0 & 0 & 0 & \mu & 0 & 0 \\ 0 & 0 & 0 & 0 & \mu & 0 \\ 0 & 0 & 0 & 0 & 0 & \mu \end{bmatrix}, \tag{3.6}$$

where the base dyads $\underline{\underline{j}}_\alpha$ are formed from the crystal base $\underline{\hat{a}}_K$. The three *independent* Lamé constants result from the classical elastic constants E, G, ν as follows:[4]

$$\lambda = \frac{E\,\nu}{(1+\nu)(1-2\nu)}, \quad \varkappa = \frac{E\,(1-\nu)}{(1+\nu)(1-2\nu)}, \quad \mu = G\,. \tag{3.7}$$

Since the relation $\mu = \frac{E}{2(1+\nu)} = \frac{1}{2}(\varkappa - \lambda)$ only holds for isotropy, the measure

$$A = \frac{2\mu}{\varkappa - \lambda} = \frac{2G}{E}(1+\nu) > 0 \tag{3.8}$$

may be used as an indicator for elastic isotropy ($A = 1$) and anistropy (else). This also shows that $\varkappa > \lambda$ must apply. In the two-dimensional case, the (fictitious) stress tensor belonging to the strain energy (3.4) can be particularly well interpreted using the lattice strain measures (2.15):

$$\underline{\underline{\widehat{T}}}_{\mathrm{e}} = \frac{\partial \hat{\phi}_{\mathrm{e}}}{\partial \underline{\underline{\widehat{E}}}_{\mathrm{e}}} = \varkappa \left\{ \varepsilon_{\mathrm{I}} \underline{\underline{\hat{A}}}_{\mathrm{I}} + \varepsilon_{\mathrm{II}} \underline{\underline{\hat{A}}}_{\mathrm{II}} \right\} + \lambda \left\{ \varepsilon_{\mathrm{I}} \underline{\underline{\hat{A}}}_{\mathrm{II}} + \varepsilon_{\mathrm{II}} \underline{\underline{\hat{A}}}_{\mathrm{I}} \right\} + \mu\, \varepsilon_{\mathrm{I,II}} \{ \underline{\hat{a}}_{\mathrm{I}} \otimes \underline{\hat{a}}_{\mathrm{II}} + \underline{\hat{a}}_{\mathrm{II}} \otimes \underline{\hat{a}}_{\mathrm{I}} \}\,. \tag{3.9}$$

Here the physical meaning of the elastic constants becomes clear: $\varkappa$ and λ describe the normal and lateral strain behavior, μ describes the shear behavior. Since the strain energy has a quadratic form, it is only admissible for *small* elastic strains and especially for *small* volume changes. Still, the formulation is applicable for arbitrarily large rotations due to the *geometrically nonlinear* strain measure. This circumstance is very important for the simulation of lattice rotations. The crystal thus has a possibility to adjust the lattice rotation (energetically) favorably.

3.3 Dislocation Energy for Cubic Primitive Crystal Lattices

To formulate the free energy of the dislocation network, dimensionless dislocation densities are introduced relating the invariants (2.26) to some characteristic value ρ_{s}:

---

[4]Here and only here in this book ν denotes Poisson's ratio (and no slip rate as everywhere else).

$$r_i = \frac{\rho_i}{\rho_s} = \frac{1}{b\rho_s}\,\hat{\underline{s}}_i \cdot \hat{\underline{\underline{\alpha}}} \cdot \hat{\underline{t}}_i = \ell\,\hat{\underline{\underline{\alpha}}} \cdot\cdot \left(\hat{\underline{s}}_i \otimes \hat{\underline{t}}_i\right)^{\mathrm{T}} , \tag{3.10}$$

where the internal (energetical) length scale $\ell = 1/(b\rho_s)$ emerges. With this, the free energy becomes the function

$$\tilde{\varrho}\psi_p = \hat{\phi}_p = \hat{\phi}_p(\hat{\underline{\underline{\alpha}}}) = \hat{\phi}_p\left(r_1(\hat{\underline{\underline{\alpha}}}), r_2(\hat{\underline{\underline{\alpha}}})\right) , \tag{3.11}$$

which—exactly as the strain energy—depends on the crystal symmetry. This symmetry should be considered (or neglected) in all constitutive relations in a *consistent* way [1].[5] Only in the special case of the quadratic form

$$\hat{\phi}_p(r_1, r_2) = \tfrac{1}{2}\,k\mu\,(r_1^2 + r_2^2) \overset{(2.27)}{=} k\mu\,\ell^2\,\|\hat{\underline{\underline{\alpha}}}\|^2 , \tag{3.12}$$

as proposed before [6], the dependence on the crystal symmetry disappears for cP-lattices and the *energetic* plastic behaviour becomes isotropic. This is no longer the case with more complicated forms. According to the argumentation from [7] a logarithmic form is considered:

$$\hat{\phi}_p(r_1, r_2) = \tfrac{1}{2}\,k\mu \sum_i \ln\left[(1 - r_i^2)^{-1}\right] = -\tfrac{1}{2}k\mu \sum_i \ln\left(1 - r_i^2\right) . \tag{3.13}$$

This form shows a quadratic behavior for small arguments. The replacement $r_i^2 \to |r_i|$ would lead to a threshold characteristic [8], which can be interpreted physically as the introduction of a nucleation energy for GNDs. From a numerical point of view, however, the handling of the absolute value function and its derivatives is problematic. In contrast, function (3.13) is continuously differentiable. In preparation for the following considerations, the derivatives of $\hat{\phi}_p$ are given:

$$\hat{\underline{\underline{X}}} := \frac{\partial \hat{\phi}_p(r_1, r_2)}{\partial \hat{\underline{\underline{\alpha}}}} = \sum_i \frac{\partial \hat{\phi}_p}{\partial r_i}\,\frac{\partial r_i(\hat{\underline{\underline{\alpha}}})}{\partial \hat{\underline{\underline{\alpha}}}} \quad \text{with} \quad \frac{\partial r_i(\hat{\underline{\underline{\alpha}}})}{\partial \hat{\underline{\underline{\alpha}}}} \overset{(3.10)}{=} \ell\,\hat{\underline{s}}_i \otimes \hat{\underline{t}}_i . \tag{3.14}$$

With the definitions (2.26) for plane strains and straight edge dislocations it follows

$$\hat{\underline{\underline{X}}} = \ell \sum_i \left(\frac{\partial \hat{\phi}_p}{\partial r_i}\,\hat{\underline{s}}_i \otimes \hat{\underline{t}}_i\right) = \ell\left(-\frac{\partial \hat{\phi}_p}{\partial r_1}\,\hat{\underline{a}}_{\mathrm{II}} + \frac{\partial \hat{\phi}_p}{\partial r_2}\,\hat{\underline{a}}_{\mathrm{I}}\right) \otimes \underline{e}_z =: \hat{\underline{x}}_z \otimes \underline{e}_z . \tag{3.15}$$

In this special case the higher order stress tensor (3.14) obviously consists only of one dyadic product, where $\hat{\underline{x}}_z$ can be interpreted as the corresponding stress *vector*.

[5] Especially with regard to feedback effects in the geometrically nonlinear theory it seems questionable to model e. g. elastic properties isotropically, but plastic properties anisotropically.

Recent work suggests *accumulated* measures for scalar dislocation densities, similar to some accumulated deformation degree [3, 4]. This can be used to model hardening effects and an (additional) history dependence. In the present work the dislocation densities as invariants of $\underline{\underline{\hat{\alpha}}}$ are state variables like the entire tensor. The physically plausible history dependence of the plastic behavior results from the integration of the differential equation (2.12) for the determination of $\underline{\underline{F}}_{\mathrm{p}}$.

3.4 Internal and External Mechanical Power

In order to specify the mechanical power, the existence of two independent tractions is postulated: the traction by stress vectors $\underline{\mathfrak{s}}$ provides the velocity $\underline{\dot{u}}(\underline{r})$ of the body and the micro traction π_i drives the plastic slip rate ν_i in the ith slip system. In addition, mass-related forces $\underline{f}$ can also occur as a result of external fields. The external power thus adopts the form

$$P_{\mathrm{ex}} = \int_A \underline{\dot{u}} \cdot \underline{\mathfrak{s}} \, \mathrm{d}A + \int_M \underline{\dot{u}} \cdot \underline{f} \, \mathrm{d}m + \int_A \sum_i \pi_i \, \nu_i \, \mathrm{d}A \,. \tag{3.16}$$

Furthermore, the validity of the following Cauchy-type relations is assumed:

$$\underline{\mathfrak{s}} = \underline{\underline{\sigma}} \cdot \underline{n} \quad , \quad \pi_i = \underline{q}_i \cdot \underline{n} \,, \tag{3.17}$$

whereby a connection is established between the external traction and the body's internal reaction with stresses $\underline{\underline{\sigma}}$ and micro stresses $\underline{q}_i$ in the ith slip system.[6] Subtracting the rate of the kinetic energy $E_{\mathrm{kin}} = \frac{1}{2} \int \underline{\dot{u}} \cdot \underline{\dot{u}} \, \mathrm{d}m$ from the external power,

$$P_{\mathrm{ex}} - \dot{E}_{\mathrm{kin}} = \int_A \underline{\dot{u}} \cdot \underline{\mathfrak{s}} \, \mathrm{d}A + \int_M \underline{\dot{u}} \cdot (\underline{f} - \underline{\ddot{u}}) \, \mathrm{d}m + \int_A \sum_i \nu_i \, \pi_i \, \mathrm{d}A = P_{\mathrm{in}} \tag{3.18}$$

results in the internal mechanical power. Applying the Gaußian integral theorem allows the transformation of the surface integral into a volume integral:

$$P_{\mathrm{in}} = \int_V \Big(\mathrm{div}(\underline{\dot{u}} \cdot \underline{\underline{\sigma}}) + \sum_i \mathrm{div}(\nu_i \underline{q}_i) + \underline{\dot{u}} \cdot \varrho \, (\underline{f} - \underline{\ddot{u}}) \Big) \, \mathrm{d}V = \int_V p_{\mathrm{in}} \, \mathrm{d}V \,, \tag{3.19}$$

[6] Since the slip system naturally has a two-dimensional structure, the micro stresses occurring in it are of dimension force/length, in contrast to the Cauchy stresses with force/surface.

from which the internal mechanical power density p_{in} can be read. Using the product rules (A.7) and the notation $\underline{v} = \dot{\underline{u}}$ it follows

$$p_{\text{in}} = \underline{\underline{\sigma}} \cdot\cdot (\underline{\nabla} \otimes \underline{v}) + \underline{v} \cdot (\underline{\underline{\sigma}} \cdot \underline{\nabla}) + \sum_i \left(\underline{q}_i \cdot (\underline{\nabla} v_i) + v_i \, (\underline{q}_i \cdot \underline{\nabla}) \right) + \underline{v} \cdot \varrho \, (\underline{f} - \ddot{\underline{u}}) \,. \tag{3.20}$$

3.5 Evaluation of the Clausius-Planck Inequality

Now the Clausius-Planck inequality is evaluated. It takes the simple form $p_{\text{in}} - \varrho \, (\dot{\psi} + \dot{\theta}\zeta) \geq 0$, where ζ denotes the specific entropy. It is important to be aware of the assumptions leading to this inequality (cf. [9, p. 517 ff.]):

- The body's mass is preserved (mass conservation).
- The total heat supply is given by the heat generation within the body minus the heat flux over the boundary.
- The total heat supply plus total external power equals the rates of internal and kinetic energy (First law).
- There is a specific entropy ζ, and a local entropy flux, entropy supply and entropy production rate.
- There is an absolute temperature θ and the local entropy flux is set equal to the local heat flux divided by θ.
- There is a specific internal energy u and the specific free energy is defined as $u - \theta\zeta$.
- The local entropy supply is equal to the heat generation divided by θ.
- The local entropy production rate g is non-negative (Second law).
- What holds for the entire body also holds for any part of it (Method of sections).

Due to the temporal constancy of the structural tensors $\hat{\underline{\underline{A}}}_K$, the free energy rate according to the form (3.1) has only three contributions, so that applies

$$\text{dissipation} := \varrho\theta g = \; p_{\text{in}} - \varrho \left(\frac{\partial \psi}{\partial \hat{\underline{\underline{E}}}_{\text{e}}} \cdot\cdot \dot{\hat{\underline{\underline{E}}}}_{\text{e}} + \frac{\partial \psi}{\partial \hat{\underline{\underline{\alpha}}}} \cdot\cdot \dot{\hat{\underline{\underline{\alpha}}}}^{\text{T}} + \frac{\partial \psi}{\partial \theta} \dot{\theta} + \dot{\theta}\zeta \right) \geq 0 \,. \tag{3.21}$$

From the terms containing the temperature rate $\dot{\theta}$, the potential relation $\zeta = -\frac{\partial \psi}{\partial \theta}$ can be inferred. Then, the partial derivatives

$$\varrho \frac{\partial \psi}{\partial \hat{\underline{\underline{E}}}_{\text{e}}} = \frac{\varrho}{\tilde{\varrho}} \frac{\partial \hat{\phi}_{\text{e}}}{\partial \hat{\underline{\underline{E}}}_{\text{e}}} = \frac{1}{j} \hat{\underline{\underline{T}}}_{\text{e}} \quad \text{and} \quad \varrho \frac{\partial \psi}{\partial \hat{\underline{\underline{\alpha}}}} = \frac{\varrho}{\tilde{\varrho}} \frac{\partial \hat{\phi}_{\text{p}}}{\partial \hat{\underline{\underline{\alpha}}}} = \frac{1}{j} \hat{\underline{\underline{X}}} \tag{3.22}$$

are inserted. Considering Formula $(2.14)_3$ and (2.9) and using $\hat{\underline{\underline{T}}}_{\text{e}} = \hat{\underline{\underline{T}}}_{\text{e}}^{\text{T}}$ it follows

$$\varrho \frac{\partial \psi}{\partial \underline{\underline{\hat{E}}}_{\mathrm{e}}} \cdot\cdot \dot{\underline{\underline{\hat{E}}}}_{\mathrm{e}} = \frac{1}{j} \underline{\underline{\widehat{T}}}_{\mathrm{e}} \cdot\cdot \underline{\underline{F}}_{\mathrm{e}}^{\mathrm{T}} \cdot \underline{\underline{D}}_{\mathrm{e}} \cdot \underline{\underline{F}}_{\mathrm{e}} = \left(\tfrac{1}{j} \underline{\underline{F}}_{\mathrm{e}} \cdot \underline{\underline{\widehat{T}}}_{\mathrm{e}} \cdot \underline{\underline{F}}_{\mathrm{e}}^{\mathrm{T}}\right) \cdot\cdot \underline{\underline{L}}^{\mathrm{T}} - \left(\tfrac{1}{j} \underline{\underline{\hat{C}}}_{\mathrm{e}} \cdot \underline{\underline{\widehat{T}}}_{\mathrm{e}}\right) \cdot\cdot \underline{\underline{\hat{L}}}_{\mathrm{p}}^{\mathrm{T}} . \tag{3.23}$$

Inserting slip system ansatz (2.7) yields for the second summand a compact form:

$$\tfrac{1}{j}\left(\underline{\underline{\hat{C}}}_{\mathrm{e}} \cdot \underline{\underline{\widehat{T}}}_{\mathrm{e}}\right) \cdot\cdot \underline{\underline{\hat{L}}}_{\mathrm{p}}^{\mathrm{T}} = \sum_i \nu_i \, \underline{\hat{s}}_i \cdot \left(\tfrac{1}{j} \underline{\underline{\hat{C}}}_{\mathrm{e}} \cdot \underline{\underline{\widehat{T}}}_{\mathrm{e}}\right) \cdot \underline{\widehat{m}}_i =: \sum_i \nu_i \, \tau_i \, , \tag{3.24}$$

where τ_i is the resolved shear stress (RSS). Next the material time derivative of $\underline{\underline{\hat{\alpha}}}$ according to Formula (2.24) is calculated which results in

$$\varrho \frac{\partial \psi}{\partial \underline{\underline{\hat{\alpha}}}} \cdot\cdot \dot{\underline{\underline{\hat{\alpha}}}}^{\mathrm{T}} = \tfrac{1}{j} \sum_i \left\{ \underline{\hat{s}}_i \cdot \underline{\underline{\hat{X}}} \cdot (\widehat{\underline{\nabla}} \nu_i \times \underline{\widehat{m}}_i) + \nu_i \left(\underline{\underline{\hat{X}}} \cdot \underline{\underline{\hat{\alpha}}}^{\mathrm{T}} + \underline{\underline{\hat{X}}}^{\mathrm{T}} \cdot \underline{\underline{\hat{\alpha}}} \right) \cdot\cdot \underline{\underline{\hat{S}}}_i^{\mathrm{T}} \right\} . \tag{3.25}$$

Factoring out the kinematic quantities ν_i and their gradients $\widehat{\underline{\nabla}} \nu_i = \underline{\underline{F}}_{\mathrm{e}}^{\mathrm{T}} \cdot \underline{\nabla} \nu_i$ with

$$\underline{\hat{s}}_i \cdot \underline{\underline{\hat{X}}} \cdot (\widehat{\underline{\nabla}} \nu_i \times \underline{\widehat{m}}_i) = \left(-\underline{\hat{s}}_i \cdot \underline{\underline{\hat{X}}} \times \underline{\widehat{m}}_i \right) \cdot \widehat{\underline{\nabla}} \nu_i = \left(\underline{\widehat{m}}_i \times \underline{\underline{\hat{X}}}^{\mathrm{T}} \cdot \underline{\hat{s}}_i \right) \cdot \underline{\underline{F}}_{\mathrm{e}}^{\mathrm{T}} \cdot \underline{\nabla} \nu_i \, , \tag{3.26}$$

two new kinetic quantities emerge: an energetic slip system back stress ς_i and a mesoscopic, distributed Peach-Koehler force $\underline{p}_i$. They are defined as follows:

$$\tfrac{1}{j} \left(\underline{\underline{\hat{X}}} \cdot \underline{\underline{\hat{\alpha}}}^{\mathrm{T}} + \underline{\underline{\hat{X}}}^{\mathrm{T}} \cdot \underline{\underline{\hat{\alpha}}} \right) \cdot\cdot \underline{\underline{\hat{S}}}_i^{\mathrm{T}} =: \varsigma_i \quad , \quad \tfrac{1}{j} \underline{\underline{F}}_{\mathrm{e}} \cdot \left(\underline{\widehat{m}}_i \times (\underline{\hat{s}}_i \cdot \underline{\underline{\hat{X}}}) \right) =: \underline{p}_i \, . \tag{3.27}$$

If now also the internal power density p_{in} according to (3.20) is inserted in Inequality (3.21), a compact and clear form of the Clausius-Planck inequality results:

$$\begin{aligned} &\left(\underline{\underline{\sigma}} \cdot \underline{\nabla} + \varrho (\underline{f} - \underline{\ddot{u}}) \right) \cdot \underline{v} \; + \; \left(\underline{\underline{\sigma}} - \tfrac{1}{j} \underline{\underline{F}}_{\mathrm{e}} \cdot \underline{\underline{\widehat{T}}}_{\mathrm{e}} \cdot \underline{\underline{F}}_{\mathrm{e}}^{\mathrm{T}} \right) \cdot\cdot (\underline{\nabla} \otimes \underline{v}) \\ + &\sum_i \left(\underline{q}_i \cdot \underline{\nabla} + (\tau_i - \varsigma_i) \right) \nu_i \; + \; \sum_i \left(\underline{q}_i - \underline{p}_i \right) \cdot (\underline{\nabla} \nu_i) \qquad \geq 0 \, . \end{aligned} \tag{3.28}$$

Both field and constitutive equations can be derived from this dissipation inequality. Under the physically evident assumption that elastic processes should be dissipation-free, the typical potential relationship for the stress tensor follows:

$$\underline{\underline{\sigma}} = \tfrac{1}{j} \underline{\underline{F}}_{\mathrm{e}} \cdot \underline{\underline{\widehat{T}}}_{\mathrm{e}} \cdot \underline{\underline{F}}_{\mathrm{e}}^{\mathrm{T}} = \tfrac{1}{j} \underline{\underline{F}}_{\mathrm{e}} \cdot \frac{\partial \hat{\phi}_{\mathrm{e}}}{\partial \underline{\underline{\hat{E}}}_{\mathrm{e}}} \cdot \underline{\underline{F}}_{\mathrm{e}}^{\mathrm{T}} . \tag{3.29}$$

Likewise, a potential relationship for the dislocation-related energetic processes in the slip system is assumed:

$$\underline{q}_i = \underline{p}_i = -\frac{1}{j}\underline{\underline{F}}_{\mathrm{e}} \cdot \left(\hat{\underline{s}}_i \cdot \frac{\partial \hat{\phi}_{\mathrm{p}}}{\partial \hat{\underline{\underline{\alpha}}}} \times \widehat{\underline{m}}_i\right) . \tag{3.30}$$

Furthermore, imposing Galilean invariance for dissipation leads to the well-known local equilibrium of forces

$$\underline{\underline{\sigma}} \cdot \underline{\nabla} + \varrho\,(\underline{f} - \underline{\ddot{u}}) = \underline{0} . \tag{3.31}$$

This leaves only the plastic slip processes as a source of entropy production (and thus irreversibility). Under the assumption that the slip processes in the different systems are independent of each other and subject to dissipation, it holds true

$$\left(\underline{q}_i \cdot \underline{\nabla} + \tau_i - \varsigma_i\right) \nu_i = \kappa_i\, \nu_i \geq 0 \quad \forall\, i \quad \text{with} \quad \kappa_i := \underline{p}_i \cdot \underline{\nabla} + \tau_i - \varsigma_i . \tag{3.32}$$

The remaining dissipation inequality $\sum \kappa_i\, \nu_i \geq 0$ can be fulfilled a priori if a relation between ν_i and effective RSS κ_i is assumed in the form of some flow rule $f_i = f(\kappa_i)$. In this way the system of constitutive equations is closed. The concrete form of the flow rules $f(\kappa_i)$ can be left open at this point. From a thermodynamic point of view, it must only be ensured that both variables take on the same sign:

$$\mathrm{sign}(\nu_i) = \mathrm{sign}(\kappa_i) \quad \forall \quad i . \tag{3.33}$$

Physically this means that plastic slip processes must always occur in the direction of the applied *effective* RSS $\kappa_i = \mathrm{div}(\underline{p}_i) + \tau_i - \varsigma_i$. Due to Relation (3.29) it now becomes obvious that τ_i follows from the true stress:

$$\hat{\underline{s}}_i \cdot \left(\frac{1}{j}\,\hat{\underline{\underline{C}}}_{\mathrm{e}} \cdot \widehat{\underline{\underline{T}}}_{\mathrm{e}}\right) \cdot \widehat{\underline{m}}_i = \tilde{\underline{s}}_i \cdot \left(\frac{1}{j}\,\widetilde{\underline{\underline{C}}} \cdot \widetilde{\underline{\underline{T}}}\right) \cdot \widetilde{\underline{m}}_i = \underline{s}_i \cdot \underline{\underline{\sigma}} \cdot \underline{m}_i =: \tau_i . \tag{3.34}$$

Consequently τ_i measures the resolved shear stress with regard to the ith *deformed* slip system in the current configuration $\mathcal{K}$.

Remark 3.1 The RSS τ_i is not a pure coefficient of the Cauchy stress tensor. This is because the deformed slip system vectors are entering, which are in general no longer unit vectors. This influence can be corrected with the normalization (2.11) and attributed—according to its nature—to the true slip rate. However, as only small elastic strains are considered here, this effect seems negligible.

Remark 3.2 Analogous to the geometrically linear theory (cf. [10]) τ_i and $\underline{q}_i$ represent the Peach-Koehler forces on dislocations in the stress field $\underline{\underline{\sigma}}$ and the higher-order stress field $\underline{\underline{\hat{X}}}$. In the geometrically nonlinear theory, the back stresses $\operatorname{div}(\underline{q}_i)$ are now supplemented by ς_i containing interactions of the slip systems (cf. p. 40).

3.6 Flow Rules

In order to establish physically meaningful phenomenological flow rules $\nu_i = f(\kappa_i)$, a precise, microstructurally based idea of the real slip processes is necessary. To make dislocations move, a critical shear stress $\tau_{\rm cr}$ must be exceeded. In addition, there is a viscous interaction between moving dislocations and crystal lattice, which limits the dislocation velocity. In addition to these experimental findings, the present study assumes that dislocations are the (only) transmitters of plastic distortion and that there is no thermo-mechanical coupling (especially no thermally activated plastic processes). Considering all these aspects, the slip processes may be divided into two (extreme) cases:

(1) Slip systems are only active if the condition $|\kappa_i| = \tau_{\rm cr}$ is fulfilled with $\tau_{\rm cr}$ denoting the critical resolved shear stress, and any viscous interaction is neglected. Then, plastic slip is rate-independent and the slip rates follow from solving the system of nonlinear equations (cf. App. A.2.2)

$$|\kappa_i(\boldsymbol{\nu})| - \tau_{\rm cr} \leq 0 \quad \forall \; i \quad \text{with} \quad \boldsymbol{\nu} = \{\nu_1, \nu_2, \ldots\}\,. \tag{3.35}$$

(2) The critical RSS is negligible compared to the viscous resistance on moving dislocations. Hence, plastic slip is rate-dependent. The viscosity $\eta_{\rm p}$ describes this resistance and is crucial for the plastic slip rates.[7] The simplest approach is

$$\nu_i = \frac{\kappa_i}{\eta_{\rm p}}\,, \tag{3.36}$$

where there is no strict case distinction elastic/plastic anymore. Moreover, all slip systems with $|\kappa_i| > 0$ are active more or less, which corresponds to $\tau_{\rm cr} = 0$. An approximate incorporation of the critical RSS is possible using the power law

$$\nu_i = \frac{\tau_{\rm cr}}{\eta_{\rm p}} \left(\frac{\kappa_i}{\tau_{\rm cr}}\right)^p \quad \forall \;\; p \in \{1, 3, 5, 7, \ldots\}\,. \tag{3.37}$$

[7] The other way round, an exact quantification of the viscosity is necessary to obtain physically meaningful results.

For $p = 1$ exactly flow rule (3.36) is obtained, flow rule (3.35) is reached asymptotically for $p \to \infty$. Flow rule (3.37) is not only mathematically advantageous over System (3.35), but can also capture a (reduced) dislocation motion below τ_{cr}. Thus, there is a smooth transition between purely elastic and elastic-plastic crystal regions. The specified flow rules can be derived from an assigned dissipation potential, which allows further information about the admissibility and validity ranges of the rules.

3.7 Potentials for Dissipation in the Slip Systems

From the thermodynamic point of view, the effective RSS $\kappa_i = \mathrm{div}(\underline{p}_i) + \tau_i - \varsigma_i$ represents a dissipative quantity. Now a more general approach than the formulation of concrete flow rules consists of postulating the existence of some dissipation potential $\mathcal{d}_i(\nu_i)$ such that κ_i follows from $\mathcal{d}_i(\nu_i)$ as

$$\kappa_i \equiv \tau_i + \mathrm{div}(\underline{p}_i) - \varsigma_i = \Lambda \frac{\partial \mathcal{d}_i(\nu_i)}{\partial \nu_i} . \tag{3.38}$$

This shows that a shear stress τ_i must be applied in the individual systems for plastic slip, supported or impeded by the back stresses $\mathrm{div}(\underline{p}_i)$ and ς_i. For the single terms in the remaining inequality $\sum \kappa_i \, \nu_i \geq 0$ it follows

$$\kappa_i \, \nu_i = \left(\underline{p}_i \cdot \underline{\nabla} + \tau_i - \varsigma_i \right) \nu_i = \Lambda \frac{\partial \mathcal{d}_i}{\partial \nu_i} \, \nu_i . \tag{3.39}$$

Under some conditions [10] the Clausius-Planck inequality is thus fulfilled a priori. Assuming accordingly that $\mathcal{d}_i$ is a homogeneous, positive function of degree n and setting $\Lambda = 1/n$, the dissipation potential equals the dissipation itself and it holds

$$\left(\underline{p}_i \cdot \underline{\nabla} + \tau_i - \varsigma_i \right) \nu_i = \frac{1}{n} \frac{\partial \mathcal{d}_i}{\partial \nu_i} \, \nu_i = \mathcal{d}_i \geq 0 . \tag{3.40}$$

This results in two extreme cases, which are discussed below. In the first case, plastic slip would be purely conservative ($\mathcal{d}_i = 0 = \kappa_i$), so the following applies:

$$\tau_i \, \nu_i = \left(\varsigma_i - \underline{p}_i \cdot \underline{\nabla} \right) \nu_i . \tag{3.41}$$

Accordingly, the entire slip system power $\tau_i \, \nu_i$ is stored energetically. In the case of a homogeneous deformation with $\underline{\underline{\hat{\alpha}}} = \underline{\underline{0}}$ there are no back stresses ς_i and $\mathrm{div}(\underline{p}_i)$, so that the unreasonable and hardly satisfiable condition $\tau_i \, \nu_i = 0$ results. From this, one can draw the well-known conclusion that plasticity must necessarily be associated with dissipation. The second (and physically possible) extreme case is purely dissipative plastic slip, so the following applies:

$$\tau_i\, \nu_i = \kappa_i\, \nu_i = d_i \geq 0 \quad \text{and} \quad \text{sign}(\nu_i) = \text{sign}(\tau_i)\,. \tag{3.42}$$

In this case the entire slip system power $\tau_i\, \nu_i$ is dissipated. In addition, slippage is then inevitably in the direction of the RSS. Plasticity due to total dissipation or total energy storage are fundamentally different mechanisms. The interplay of both mechanisms influences the (theoretical) material behavior to a high degree.

To close the theory, a specific form for all potentials d_i is necessary. For this, equal conditions in all slip systems are assumed, such that only one function $d(\nu_i)$ is needed. In this paper three homogeneous functions of different degrees are considered:

$$d(\nu_i) = \tau_{\text{cr}} |\nu_i| \quad \text{as well as} \quad d(\nu_i) = \eta_{\text{p}}\, \nu_i^2\,. \tag{3.43}$$

The power law (3.37) follows for a dissipation potential of the form

$$d(\nu_i) = \tau_{\text{cr}}\, \nu_i \left(\frac{\nu_i}{\bar{\nu}}\right)^{\frac{1}{p}} \quad \text{with} \quad \bar{\nu} = \frac{\tau_{\text{cr}}}{\eta_{\text{p}}}\,, \tag{3.44}$$

which couples the dissipative mechanisms in a nonlinear way. With Relation (3.38) and $\Lambda = \frac{p}{1+p}$, flow rule (3.37) follows. The case $p = 1$ yields the quadratic form $(3.43)_2$. For $p = 3, 5, 7, \ldots$ the function adapts itself more and more to the absolute value function $(3.43)_1$, which is only reached asymptotically as $p \to \infty$ (cf. Fig. 3.1). The essential (numerical) advantage compared to Form $(3.43)_1$ is that Potential (3.44) is *smooth* and hence continuously differentiable.

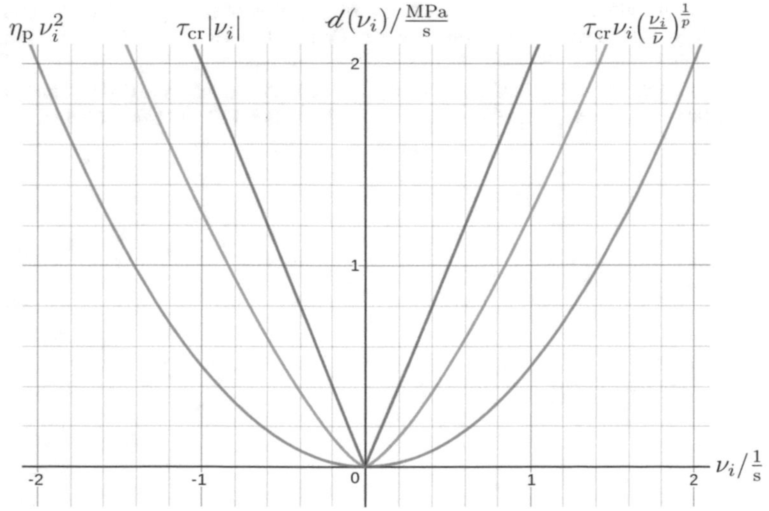

Fig. 3.1 Illustration of three different forms of the dissipation potential with the parameter values $\tau_{\text{cr}} = 2$ MPa, $\eta_{\text{p}} = 0.5$ MPa s, $\bar{\nu} = \tau_{\text{cr}}/\eta_{\text{p}} = 4\frac{1}{s}$ and $p = 3$

Remark 3.3 Instead of the slip rates ν_i (of the intermediate configuration) the dissipation d can also be formulated as a function of the true slip rates. This, however, results in an additional dependence of d on the elastic distortion, which has to be considered consistently from assumption (2.8) onwards and makes the following derivatives much more complicated.

References

1. Gurtin, M.E.: A gradient theory of single-crystal viscoplasticity that accounts for geometrically necessary dislocations. J. Mech. Phys. Solids **50**(1), 5–32 (2002)
2. Ihlemann, J.: Beobachterkonzepte und Darstellungsformen der nichtlinearen Kontinuumsmechanik. VDI-Verlag, Düsseldorf (2014)
3. Gurtin, M.E.: A finite-deformation, gradient theory of single-crystal plasticity with free energy dependent on densities of geometrically necessary dislocations. Int. J. Plast. **24**(4), 702–725 (2008)
4. Gurtin, M.E.: A finite-deformation, gradient theory of single-crystal plasticity with free energy dependent on the accumulation of geometrically necessary dislocations. Int. J. Plast. **26**(8), 1073–1096 (2010)
5. Haupt, P.: Continuum Mechanics and Theory of Materials, vol. 2. Springer, Berlin (2002)
6. Gurtin, M.E., Anand, L.: A theory of strain-gradient plasticity for isotropic, plastically irrotational materials. Part II: Finite deformations. Int. J. Plast. **21**(12), 2297–2318 (2005)
7. Berdichevsky, V.L., Le, K.C.: Dislocation nucleation and work hardening in anti-plane constrained shear. Continuum Mech. Thermodyn. **18**(7–8), 455–467 (2007)
8. Le, K.C., Sembiring, P.: Analytical solution of plane constrained shear problem for single crystals within continuum dislocation theory. Arch. Appl. Mech. **78**(8), 587–597 (2008)
9. Ottosen, N.S., Ristinmaa, M.: The Mechanics of Constitutive Modeling. Elsevier, Amsterdam (2005)
10. Silbermann, C.B., Ihlemann, J.: Geometrically linear continuum theory of dislocations revisited from a thermodynamical perspective. Arch. Appl. Mech. **88**(1–2), 141–173 (2017)

Chapter 4
Special Cases Included in the Theory

Abstract The general geometrically nonlinear continuum dislocation theory on the basis of [1] presented in the previous chapter includes four important special cases, which are briefly described in this chapter. This also involves a rigorous treatment of the important case of single slip, and the geometrical linearization. Thereby, some subtle but significant differences to geometrically *linear* continuum dislocation theory are discovered.

4.1 Single Crystal Elasticity

$\underline{\underline{F}}_{\mathrm{p}}(t) = \underline{\underline{I}}$ or $\boldsymbol{\nu}(t) = \mathbf{0}\ \forall t$ yields anisotropic, hyperelastic elasticity for cubic single crystals under large deformations. The material behavior is therefore geometrically nonlinear. By using the quadratic strain energy (3.4) the *physically* linear stress-strain relationship is obtained

$$\widehat{\underline{\underline{T}}}_{\mathrm{e}} = \frac{\partial \tilde{\rho}\psi}{\partial \hat{\underline{\underline{E}}}_{\mathrm{e}}} = \widehat{\underline{\underline{\underline{\underline{K}}}}} \cdot\cdot\, \hat{\underline{\underline{E}}}_{\mathrm{e}} \quad \underset{\underline{\underline{F}}=\underline{\underline{F}}_{\mathrm{e}}}{\rightarrow} \quad \widetilde{\underline{\underline{T}}} = \widetilde{\underline{\underline{\underline{\underline{K}}}}} \cdot\cdot\, \underline{\underline{E}}$$

with the second Piola-Kirchhoff stress tensor $\widetilde{\underline{\underline{T}}}$ and the *constant* fourth-order elasticity tensor $\widetilde{\underline{\underline{\underline{\underline{K}}}}} \equiv \widehat{\underline{\underline{\underline{\underline{K}}}}}$ according to the coefficient matrix (3.5).

4.2 Single Crystal Plasticity

The special cases of compatible plastic distortion due to $\mathrm{Curl}(\underline{\underline{F}}_{\mathrm{p}}) = \underline{\underline{0}}$ or vanishing free energy ψ_{p} of the dislocation network yields the following standard model of single crystal elasto(visco)plasticity.

C. B. Silbermann et al., *Introduction to Geometrically Nonlinear Continuum Dislocation Theory*, SpringerBriefs in Continuum Mechanics,
https://doi.org/10.1007/978-3-030-63696-8_4

Kinematics:	$\underline{\underline{F}}_{\rm e} = \underline{\underline{F}} \cdot \underline{\underline{F}}_{\rm p}^{-1}\ ,\qquad \hat{\underline{\underline{C}}}_{\rm e} = \underline{\underline{F}}_{\rm e}^{\rm T} \cdot \underline{\underline{F}}_{\rm e}\ ,\quad \hat{\underline{\underline{E}}}_{\rm e} = \frac{1}{2}(\hat{\underline{\underline{C}}}_{\rm e} - \underline{\underline{I}})\ ,$ $\dot{\underline{\underline{F}}}_{\rm p} = \hat{\underline{\underline{L}}}_{\rm p} \cdot \underline{\underline{F}}_{\rm p} = \sum_i \left(\nu_i\, \hat{\underline{s}}_i \otimes \widehat{\underline{m}}_i\right) \cdot \underline{\underline{F}}_{\rm p}\ .$
Kinetics:	$\widehat{\underline{\underline{T}}}_{\rm e} = \widehat{\underline{\underline{\underline{\underline{K}}}}} \cdot\cdot\, \hat{\underline{\underline{E}}}_{\rm e}\ ,\quad \tau_i = \hat{\underline{s}}_i \cdot (\frac{1}{j}\hat{\underline{\underline{C}}}_{\rm e} \cdot \widehat{\underline{\underline{T}}}_{\rm e}) \cdot \widehat{\underline{m}}_i\ ,\quad \nu_i = f(\tau_i(\boldsymbol{\nu}))\ .$

This model has been proposed long before [2]. The slip rates ν_i represent inner variables, for which merely some evolution laws in the form of the flow rules $f(\tau_i(\boldsymbol{\nu}))$ need to be specified. Depending on the form of $f(\tau_i(\boldsymbol{\nu}))$, rate-dependent (viscoplastic) or rate-independent (plastic) material behavior can be modeled.

4.3 Single Slip

Another special case occurs if only one slip system is active: by virtue of $\hat{\underline{\underline{L}}}_{\rm p} = \nu\, \hat{\underline{s}} \otimes \widehat{\underline{m}}$ and

$$\exp\left(\left(\Delta t\, \hat{\underline{\underline{L}}}_{\rm p}\right)\right) = \exp\left(\left(\nu(t{+}\Delta t)\, \Delta t\ \hat{\underline{s}} \otimes \widehat{\underline{m}}\right)\right) = \underline{\underline{I}} + \nu(t{+}\Delta t)\, \Delta t\, \hat{\underline{s}} \otimes \widehat{\underline{m}} \tag{4.1}$$

and considering Rule (2.4), the tensor differential equation (2.12) has the solution

$$\underline{\underline{F}}_{\rm p}(t{+}\Delta t) = \left(\underline{\underline{I}} + \xi(t{+}\Delta t)\, \hat{\underline{s}} \otimes \widehat{\underline{m}}\right) \cdot \underline{\underline{F}}_{\rm p}(t) = \underline{\underline{F}}_{\rm p}(t) + \xi(t{+}\Delta t)\, \hat{\underline{s}} \otimes \widetilde{\underline{m}}(t)\ , \tag{4.2}$$

where $\xi(t{+}\Delta t) := \nu(t{+}\Delta t)\Delta t$ is the plastic increment. Evaluating this equation iteratively under initial condition $\underline{\underline{F}}_{\rm p}(t_0) = {}^0\underline{\underline{F}}_{\rm p}$ for $t_n = t_0 + n\Delta t$ $(n = 1, 2, \ldots, N)$ instants of time and exploiting $\hat{\underline{s}} \cdot \widehat{\underline{m}} = 0$, the explicit update formula

$$\underline{\underline{F}}_{\rm p}(t_{N+1}) = \left(\underline{\underline{I}} + \sum_{n=1}^{N} \xi(t_n)\, \hat{\underline{s}} \otimes \widehat{\underline{m}}\right) \cdot \underline{\underline{F}}_{\rm p}(t_0) = {}^0\underline{\underline{F}}_{\rm p} + \sum_{n=1}^{N} \xi(t_n)\, \hat{\underline{s}} \otimes \widetilde{\underline{m}}(t_0) \tag{4.3}$$

is obtained. This solution is plausible for two reasons: First, tensors are added that have the same transformation properties (mappings from the reference configuration onto the lattice space); second, the current value $\underline{\underline{F}}_{\rm p}(t_{N+1})$ depends on the history of $\underline{\underline{F}}_{\rm p}$ (or ν). The frequently used *ad-hoc* assumption [3–7]

$$\underline{\underline{F}}_{\rm p}(t) = \underline{\underline{I}} + \beta_{\rm p}(t)\, \hat{\underline{s}} \otimes \widehat{\underline{m}} \qquad \text{with} \qquad \dot{\underline{\underline{F}}}_{\rm p}(t) = \dot{\beta}_{\rm p}(t)\, \hat{\underline{s}} \otimes \widehat{\underline{m}} \tag{4.4}$$

contradicts the afore-mentioned criteria: the history dependence is lost and $\underline{\underline{F}}_{\rm p}(t)$ does not map from the reference configuration onto the lattice space as it should. In order to do so, the slip system normal in Eq. (4.4) must be a vector of the reference configuration $\widetilde{\mathcal{K}}$ ($\underline{\widehat{m}} \rightarrow \underline{\widetilde{m}}$), whereupon $\underline{\widetilde{m}}$ would not be constant anymore.[1] Moreover, tensors cannot be elements of different spaces/configurations [8]: $\underline{\hat{s}}$, $\underline{\widehat{m}}$ are neither vectors of $\widetilde{\mathcal{K}}$ nor $\mathcal{K}$, they clearly belong to the lattice space $\widehat{\mathcal{K}}$. As a solution of a differential equation, $\underline{\underline{F}}_{\mathrm{p}}$ must depend not only on the current slip rates, but on their complete history [6]. This history dependence can hardly be captured with an algebraic equation at the deformation level. Consequently, the *ad-hoc* ansatz (4.4) is *no* solution of the differential equation (2.12) in general. The only exception is under the IC $\underline{\underline{F}}_{\mathrm{p}}(t_0) = \underline{\underline{I}}$, where $\beta_{\mathrm{p}} = \sum \xi(t_n)$ would have to be treated like a history variable. However, this is not the case [4, 5, 9].[2]

4.4 Geometrical Linearization of the Theory

The presented continuum dislocation theory for large deformations shall now be linearized geometrically. By the following comparison with the geometrically linear theory [10], the plausibility can be evaluated.

For small (elastic and plastic) deformations, the following applies:

$$\underline{\underline{F}}_{\mathrm{e}} \rightarrow \underline{\underline{I}} \ , \quad \underline{\underline{F}}_{\mathrm{p}} \rightarrow \underline{\underline{I}} \ , \quad \underline{\underline{F}} \rightarrow \underline{\underline{I}} \ , \quad j = I_3\big(\underline{\underline{F}}\big) \rightarrow 1 \ . \tag{4.5}$$

Hence, the crystal lattice becomes static, i. e. the lattice vectors remain constant:

$$\underline{a}_K = \underline{\hat{a}}_K = \underline{\tilde{a}}_K \ , \quad \underline{s}_i = \underline{\hat{s}}_i = \underline{\tilde{s}}_i \ , \quad \underline{m}_i = \underline{\widehat{m}}_i = \underline{\widetilde{m}}_i \ . \tag{4.6}$$

At first, no slip system approach is presumed, but it is assumed generally

$$\underline{\underline{F}}_{\mathrm{p}} = \underline{\underline{\beta}}_{\mathrm{p}} + \underline{\underline{I}} \quad \text{with} \quad \underline{\underline{\beta}}_{\mathrm{p}} \rightarrow \underline{\underline{0}} \ . \tag{4.7}$$

For the dislocation density tensor according to Definition (2.18) it follows

$$\underline{\underline{\hat{\alpha}}} = \mathrm{Curl}(\underline{\underline{F}}_{\mathrm{p}}) \cdot \underline{\underline{F}}_{\mathrm{p}}^{\mathrm{T}} = \mathrm{Curl}(\underline{\underline{\beta}}_{\mathrm{p}}) \cdot (\underline{\underline{\beta}}_{\mathrm{p}} + \underline{\underline{I}})^{\mathrm{T}} \approx \mathrm{Curl}(\underline{\underline{\beta}}_{\mathrm{p}}) = \underline{\underline{\alpha}} \ , \tag{4.8}$$

which matches geometrically linear theory exactly. Next, the rate of the dislocation density tensor is linearized according to the Formula (2.20)[3]:

$$\dot{\underline{\underline{\hat{\alpha}}}} = \left(\text{-}\underline{\underline{\beta}}_{\mathrm{p}} \times \underline{\widetilde{\nabla}}\right) \cdot \underline{\underline{\dot{\beta}}}_{\mathrm{p}}^{\mathrm{T}} + \left(\text{-}\underline{\underline{\beta}}_{\mathrm{p}} \times \underline{\widetilde{\nabla}}\right)^{\cdot} \cdot (\underline{\underline{\beta}}_{\mathrm{p}} + \underline{\underline{I}})^{\mathrm{T}} \approx \underline{\underline{\alpha}} \cdot \underline{\underline{\dot{\beta}}}_{\mathrm{p}}^{\mathrm{T}} + \mathrm{Curl}(\underline{\underline{\dot{\beta}}}_{\mathrm{p}}) \ . \tag{4.9}$$

---

[1] The notation has been adapted in order to point up that $\underline{\hat{s}} \otimes \underline{\widehat{m}}$ is assumed constant.

[2] Still, the approach of [4] makes sense insofar as it allows *analytical* considerations on the convexity of the energy and statements on the stability of the solution become possible at all.

[3] This represents the mathematically correct order: first differentiation, then linearization.

This reveals a fundamental difference to geometrically linear theory, where it is assumed $\dot{\underline{\underline{\alpha}}} = \mathrm{curl}(\dot{\underline{\underline{\beta}}}_\mathrm{p})$ according to Formula (4.8). As the linearization now shows, this assumption seems only valid if the term $\underline{\underline{\alpha}} \cdot \dot{\underline{\underline{\beta}}}_\mathrm{p}^\mathrm{T}$ is negligible. In general, this is only fulfilled for $\|\underline{\underline{\alpha}}\| \to 0$, i.e. the plastic incompatibility and thus also the dislocation density is *small*. This result considerably restricts the applicability of the linear theory (cf. e.g. [10, and the references therein]). However, there is an important exception for the special case of single-slip, being considered now.

For one active slip system $\underline{\underline{S}} = \underline{s} \otimes \underline{m}$ with $\underline{\underline{\beta}}_\mathrm{p} = \beta_\mathrm{p}\, \underline{\underline{S}}$ and $\underline{\underline{\alpha}} = \mathrm{Curl}(\underline{\underline{\beta}}_\mathrm{p})$ it follows indeed by implication

$$\underline{\underline{\alpha}} \cdot \dot{\underline{\underline{\beta}}}_\mathrm{p}^\mathrm{T} = (\underline{s} \otimes \underline{\nabla}\beta_\mathrm{p} \times \underline{m}) \cdot (\dot{\beta}_\mathrm{p}\, \underline{s} \otimes \underline{m})^\mathrm{T} = \dot{\beta}_\mathrm{p}\, \underline{s} \otimes (\underline{\nabla}\beta_\mathrm{p}) \cdot (\underline{m} \times \underline{m}) \otimes \underline{s} = \underline{\underline{0}}\,. \tag{4.10}$$

As for single-slip holds $\underline{\underline{\alpha}} \cdot \underline{\underline{S}}^\mathrm{T} = \underline{\underline{S}} \cdot \underline{\underline{\alpha}} = \underline{\underline{0}}$ it follows for the linearization of Eq. (2.24)

$$\dot{\hat{\underline{\underline{\alpha}}}} = \hat{\underline{s}} \otimes \widehat{\underline{\nabla}}\nu \times \widehat{\underline{m}} \approx \underline{s} \otimes \underline{\nabla}\dot{\beta}_\mathrm{p} \times \underline{m} = \mathrm{curl}\left(\dot{\underline{\underline{\beta}}}_\mathrm{p}\right) \tag{4.11}$$

completely analogous to geometrically linear theory. Accordingly, the numerical results presented before [10] are also valid for *large* dislocation densities. Furthermore, it follows that the energetic back stresses ς $(3.27)_1$ vanish. Finally, Formula (4.3) is linearized inserting expression (4.7):

$$\underline{\underline{\beta}}_\mathrm{p}(t{+}\Delta t) - \underline{\underline{\beta}}_\mathrm{p}(t) = \nu(t{+}\Delta t)\, \Delta t\, \underline{s} \otimes \underline{m} \quad \Leftrightarrow \quad \dot{\underline{\underline{\beta}}}_\mathrm{p} = \nu(t)\, \underline{s} \otimes \underline{m}\,. \tag{4.12}$$

Integration over the elapsed time $\mathscr{t} \le t$ yields the plastic distortion as[4]

$$\underline{\underline{\beta}}_\mathrm{p}(t) = \int\limits_{\mathscr{t}=0}^{\mathscr{t}=t} \nu(\mathscr{t})\, \mathrm{d}\mathscr{t}\;\, \underline{s} \otimes \underline{m} =: \beta_\mathrm{p}(\mathscr{t})\, \underline{s} \otimes \underline{m}\,, \tag{4.13}$$

which at first glance resembles the usual approach for single-slip in linear theory. The fundamental difference, however, is that $\beta_\mathrm{p}(\mathscr{t})$ $(\mathscr{t} \le t)$ is determined by a history integral, similar to the accumulated degree of deformation in phenomenological viscoplasticity (cf. e.g. [11]). This seems absolutely evident, since irreversible plastic distortion leads to a memory behavior of the material. However, by replacing $\beta_\mathrm{p}(\mathscr{t})$ in Eq. (4.13) with $\beta_\mathrm{p}(t)$ at the current time t, any history is lost. If, in addition, zero dissipation is assumed and the plastic slip $\beta_\mathrm{p}(t)$ is determined by pure energy minimization [12–14], the characteristic nature of plasticity might get lost. This criticism is directly transferable to the *algebraic* approach for multi-slip [9]. Conversely, the

[4]Note that only for $\nu = \mathrm{const.}$ the integration can be omitted and $\beta_\mathrm{p}(\mathscr{t}) = \beta_\mathrm{p}(t) = \nu\, t$.

origin of some associated problems [15] seems hence identified. The conclusion is obvious: A slip system approach generally only makes sense on the *rate* level, which leads to an evolution equation for the plastic distortion.

References

1. Gurtin, M.E.: A gradient theory of single-crystal viscoplasticity that accounts for geometrically necessary dislocations. J. Mech. Phys. Solids **50**(1), 5–32 (2002)
2. Asaro, R.J., Rice, J.R.: Strain localization in ductile single crystals. J. Mech. Phys. Solids **25**(5), 309–338 (1977)
3. Bortoloni, L., Cermelli, P.: Dislocation Patterns and Work-Hardening in Crystalline Plasticity. J. Elast. **76**(2), 113–138 (2004)
4. Koster, M., Le, K.C., Nguyen, B.D.: Formation of grain boundaries in ductile single crystals at finite plastic deformations. Int. J. Plast. **69**, 134–151 (2015)
5. Le, K.C., Günther, C.: Nonlinear continuum dislocation theory revisited. Int. J. Plast. **53**, 164–178 (2014)
6. Levitas, V.I., Javanbakht, M.: Thermodynamically consistent phase field approach to dislocation evolution at small and large strains. J. Mech. Phys. Solids **82**, 345–366 (2015)
7. Conti, S., Hackl, K. (eds.): Analysis and Computation of Microstructure in Finite Plasticity. Springer, Berlin (2015)
8. Gurtin, M.E., Reddy, B.D.: Some issues associated with the intermediate space in single-crystal plasticity. J. Mech. Phys. Solids **95**, 230–238 (2016)
9. Le, K.C.: Three-dimensional continuum dislocation theory. Int. J. Plast. **76**, 213–230 (2016)
10. Silbermann, C.B., Ihlemann, J.: Geometrically linear continuum theory of dislocations revisited from a thermodynamical perspective. Arch. Appl. Mech. **88**(1–2), 141–173 (2017)
11. Shutov, A.V., Kuprin, C., Ihlemann, J., Wagner, M.F.X., Silbermann, C.: Experimentelle Untersuchung und numerische Simulation des inkrementellen Umformverhaltens von Stahl 42CrMo4. Materialwiss. Werkstofftech. **41**(9), 765–775 (2010)
12. Le, K.C., Nguyen, Q.S.: Polygonization as low energy dislocation structure. Continuum Mech. Thermodyn. **22**(4), 291–298 (2010)
13. Le, K.C., Sembiring, P.: Analytical solution of plane constrained shear problem for single crystals within continuum dislocation theory. Arch. Appl. Mech. **78**(8), 587–597 (2008)
14. Le, K.C., Sembiring, P.: Plane constrained uniaxial extension of a single crystal strip. Int. J. Plast. **25**(10), 1950–1969 (2009)
15. Silbermann, C.B., Ihlemann, J.: Kinematic assumptions and their consequences on the structure of field equations in continuum dislocation theory. IOP Conf. Ser. Mater. Sci. Eng. **118**, 012,034+ (2016)

Chapter 5
Variational Formulation of the Theory

Abstract In this chapter a variational formulation of the geometrically nonlinear continuum dislocation theory is developed based on the principle of virtual power [1]. Inserting special virtual motions, the balance equations of linear and angular momentum are derived. For the case of a continuously dislocated, plane, cubic primitive single crystal with two active slip systems, the governing integral equations are transformed to matrix notation.

5.1 Principle of Virtual Power for Dissipative Mechanical Systems

A *dissipative* thermomechanical system is considered, to which the power P_{ex} and the heat flux Q are supplied. This causes a change of the inner energy U and the kinetic energy E_{kin}. The global power balance is then [1]:

$$P_{\text{ex}} - \dot{E}_{\text{kin}} = \dot{U} - Q\,, \quad \text{where} \quad \dot{U} - Q \equiv \dot{\mathcal{F}} + \mathcal{D} - Q^{\text{AF}} \tag{5.1}$$

essentially follows from the Legendre-transformation from *inner* to *free* energy. The power balance (5.1) corresponds exactly to the 1st law of thermodynamics (cf. p. 20). The principle of *virtual* power according to [1] provides a fundamental statement beyond that. It is characteristic that the rates of the primary fields are varied. Since only the *mechanical* subproblem is of interest here, the variation of those contributions that only refer to the *thermal* subproblem (Q or Q^{AF}) are not considered further.

Theorem 5.1 (Principle of virtual power for dissipative mechanical systems) *In dissipative mechanical systems, the actual primary fields among all admissible primary fields are determined in such a way that the power balance* (5.1) *also applies to the* virtual *quantities, i. e. for the mechanical subproblem:*

$$\delta\left(P_{ex} - \dot{E}_{kin}\right) = \delta\left(\dot{\mathcal{F}} + \mathcal{D}\right)\,. \tag{5.2}$$

C. B. Silbermann et al., *Introduction to Geometrically Nonlinear Continuum Dislocation Theory*, SpringerBriefs in Continuum Mechanics,
https://doi.org/10.1007/978-3-030-63696-8_5

Besides the equality of these virtual powers, the same statement re-arranged to

$$\delta\left(P_{ex} - \dot{E}_{kin} - \dot{\mathcal{F}} - \mathcal{D}\right) = 0 \tag{5.3}$$

can be read as a stationarity condition: The actual primary fields lead to a stationary value of $P_{ex} - \dot{E}_{kin} - \dot{\mathcal{F}} - \mathcal{D}$*, which characterizes mechanical equilibrium.*[1]

For the special case that the free energy remains constant and no entropy is produced ($\dot{\mathcal{F}} = \mathcal{D} = 0$), the well-known *principle of Jourdain* results: $\delta P_{ex} = \delta \dot{E}_{kin}$.

5.2 General Tensor Formulation

The principle of virtual power is valid for any (mechanical) material behaviour. It is now applied to geometrically nonlinear continuum dislocation theory. Therefore arbitrary, independent and kinematically admissible virtual velocities $\delta\underline{\dot{u}} = \delta\underline{v}$ and slip rates $\delta\nu_i$ are introduced. In the context of continuum dislocation theory, Eq. (5.2) can be specified as *principle of virtual velocities and slip rates*.

Variation of $P_{ex} - \dot{E}_{kin}$ (3.18) yields the virtual internal power

$$\delta\left(P_{ex} - \dot{E}_{kin}\right) = \int\limits_A \underline{s} \cdot \delta\underline{\dot{u}}\, dA + \int\limits_M (\underline{f} - \underline{\ddot{u}}) \cdot \delta\underline{\dot{u}}\, dm + \int\limits_A \sum_i \pi_i\, \delta\nu_i\, dA\,. \tag{5.4}$$

Here $\underline{s}$ and π_i denote the traction vector and some micro traction in the i-th slip system, prescribed at the boundary. Next, the free energy rate (3.2) is given, assuming the Formula (3.3) and omitting the purely thermal contributions:

$$\dot{\mathcal{F}} = \left(\int\limits_M \psi\, dm\right)^{\cdot} \underset{\dot{m}=0}{=} \int\limits_M \dot{\psi}\, dm = \int\limits_V \varrho\dot{\psi}\, dV = \int\limits_V \varrho\left(\frac{\partial\psi}{\partial\underline{\underline{\hat{E}}}_e} \cdot\cdot\, \dot{\underline{\underline{\hat{E}}}}_e + \frac{\partial\psi}{\partial\underline{\underline{\hat{\alpha}}}} \cdot\cdot\, \dot{\underline{\underline{\hat{\alpha}}}}^{T}\right) dV\,. \tag{5.5}$$

Exploiting the Relations (3.23)–(3.27) the variation of the free energy is obtained as

$$\delta\dot{\mathcal{F}} = \int\limits_V \left(\underline{\underline{P}} \cdot\cdot\, (\underline{\nabla} \otimes \delta\underline{\dot{u}}) - \sum_i (\tau_i - \varsigma_i)\, \delta\nu_i + \sum_i \underline{p}_i \cdot (\underline{\nabla}\, \delta\nu_i)\right) dV\,, \tag{5.6}$$

[1]The point $P_{ex} - \dot{E}_{kin} - \dot{\mathcal{F}} - \mathcal{D} = 0$ is thus stationary. This restricts the possibilities offered by the first law. Furthermore, with the dissipation $\mathcal{D} \geq 0$ the second law is fulfilled as well.

where the abbreviation $\underline{\underline{P}} = \frac{1}{j}\underline{\underline{F}}_{\mathrm{e}} \cdot \widehat{\underline{\underline{T}}}_{\mathrm{e}} \cdot \underline{\underline{F}}_{\mathrm{e}}^{\mathrm{T}}$ was introduced, and it was utilized that the order of variation and differentiation is commutable. For the dissipation, the form $\sum \kappa_i \nu_i$ is assumed, where the dissipative shear stresses $\kappa_i = \Lambda\, \partial d / \partial \nu_i$ are derived from potential d. The (virtual) dissipation thus is

$$\mathcal{D} = \int\limits_V \sum_i \kappa_i \nu_i \,\mathrm{d}V \quad \rightarrow \quad \delta\mathcal{D} = \int\limits_V \sum_i \kappa_i \,\delta\nu_i \,\mathrm{d}V = \int\limits_V \sum_i \Lambda \frac{\partial d\,(\boldsymbol{\nu})}{\partial \nu_i} \,\delta\nu_i \,\mathrm{d}V\,. \tag{5.7}$$

Considering all this, Principle (5.2) leads to the following *weak* formulation of mechanical equilibrium for a continuously dislocated crystal:

$$\sum_i \int\limits_V \left(\underline{p}_i \cdot (\underline{\nabla}\, \delta\nu_i) - \left(\tau_i - \varsigma_i - \Lambda \frac{\partial d}{\partial \nu_i} \right) \delta\nu_i \right) \mathrm{d}V + \int\limits_V \left(\underline{\underline{P}} \cdot\cdot (\underline{\nabla} \otimes \delta\underline{\upsilon}) - \varrho\,(\underline{f} - \underline{\ddot{u}}) \cdot \delta\underline{\upsilon} \right) \mathrm{d}V = \int\limits_A \left(\underline{s} \cdot \delta\underline{\upsilon} + \sum_i \pi_i \,\delta\nu_i \right) \mathrm{d}A\,. \tag{5.8}$$

This is the principle of virtual velocities and slip rates. By looking at special virtual motions, conservation laws can be obtained: For a virtual rigid body *translation* $\delta\nu_i = 0\;\forall\, i$ and $\delta\underline{\upsilon} = \delta\underline{c} = \text{const.}$ applies. This turns Eq. (5.8) into

$$\left(\int\limits_V \varrho\,(\underline{f} - \underline{\ddot{u}})\,\mathrm{d}V + \int\limits_A \underline{s}\,\mathrm{d}A \right) \cdot \delta\underline{c} = 0\,. \tag{5.9}$$

According to the fundamental theorem of variational calculus, the expression in parentheses must vanish, which yields the well-known principle of linear momentum in the global form

$$\int\limits_V \varrho\,\underline{f}\,\mathrm{d}V + \int\limits_A \underline{s}\,\mathrm{d}A = \int\limits_V \varrho\,\underline{\ddot{u}}\,\mathrm{d}V\,. \tag{5.10}$$

For a virtual rigid body *rotation* $\delta\nu_i = 0\;\forall\, i$ and $\delta\underline{\upsilon} = \delta\underline{\omega} \times \underline{r}$ applies with the angular velocity $\delta\underline{\omega} = \text{const.}$ This results in a constant virtual velocity gradient $\underline{\nabla} \otimes \delta\underline{\upsilon} = -\underline{\nabla} \otimes \underline{r} \times \delta\underline{\omega} = \underline{\underline{\underline{\epsilon}}} \cdot \delta\underline{\omega}$ (cf. Appendix A.1.1). Insertion into Eq. (5.8) yields

$$\delta\underline{\omega} \cdot \left(\int\limits_V \underline{r} \times \varrho\,(\underline{f} - \underline{\ddot{u}})\,\mathrm{d}V + \int\limits_A \underline{r} \times \underline{s}\,\mathrm{d}A - \int\limits_V \underline{\underline{P}} \cdot\cdot \underline{\underline{\underline{\epsilon}}}\,\mathrm{d}V \right) = 0\,. \tag{5.11}$$

According to the fundamental theorem of variational calculus, the expression in parentheses must vanish, which is fulfilled with the following global balance equations:

$$\int_V \underline{r} \times \varrho \underline{f} \, \mathrm{d}V + \int_A \underline{r} \times \underline{\mathfrak{s}} \, \mathrm{d}A = \int_V \underline{r} \times \varrho \ddot{\underline{u}} \, \mathrm{d}V \quad \text{and} \quad \int_V \underline{\underline{P}} \cdot\cdot \underline{\underline{\underline{\epsilon}}} \, \mathrm{d}V = \underline{0} \, . \tag{5.12}$$

The first balance corresponds to the well-known principle of angular momentum. From the second balance follows a local statement in the form of the symmetry condition $\underline{\underline{P}} = \underline{\underline{P}}^{\mathrm{T}}$, so that $\underline{\underline{P}} \cdot\cdot \underline{\underline{\underline{\epsilon}}} = \underline{0}$. Provided this and taking into account Eq. (5.10), the global balance $(5.12)_1$ is fulfilled identically. Accordingly, the principle of angular momentum essentially results in condition $(5.12)_2$.

In order to arrive at the *strong* formulation, the gradients of the primary fields $\underline{v}$ and ν_i are eliminated integrating by parts. Using the product rules (A.7) gives

$$\begin{aligned}
\int_V \underline{p}_i \cdot (\underline{\nabla} \, \delta\nu_i) \, \mathrm{d}V &= \int_V \Big((\delta\nu_i \, \underline{p}_i) \cdot \underline{\nabla} - \delta\nu_i \, (\underline{p}_i \cdot \underline{\nabla}) \Big) \, \mathrm{d}V \\
&= \oint_A \delta\nu_i \, \underline{p}_i \cdot \underline{n} \, \mathrm{d}A - \int_V \delta\nu_i \, (\underline{p}_i \cdot \underline{\nabla}) \, \mathrm{d}V , \\
\int_V \underline{\underline{P}} \cdot\cdot (\underline{\nabla} \otimes \delta\underline{v}) \, \mathrm{d}V &= \int_V \big((\delta\underline{v} \cdot \underline{\underline{P}}) \cdot \underline{\nabla} - \delta\underline{v} \cdot (\underline{\underline{P}} \cdot \underline{\nabla}) \big) \, \mathrm{d}V \\
&= \oint_A \delta\underline{v} \cdot \underline{\underline{P}} \cdot \underline{n} \, \mathrm{d}A - \int_V \delta\underline{v} \cdot (\underline{\underline{P}} \cdot \underline{\nabla}) \, \mathrm{d}V .
\end{aligned}$$

Here, the Gaußian integral theorem was used to partially transform volume integrals into surface integrals (with $A = \partial V$). The principle of the virtual velocities and slip rates within nonlinear continuum dislocation theory is thus

$$\begin{aligned}
&\sum_i \int_V \left(\Big(\underline{p}_i \cdot \underline{\nabla} \Big) + (\tau_i - \varsigma_i) - \Lambda \frac{\partial d}{\partial \nu_i} \right) \delta\nu_i \, \mathrm{d}V + \int_V \left(\Big(\underline{\underline{P}} \cdot \underline{\nabla} \Big) + \varrho (\underline{f} - \ddot{\underline{u}}) \right) \cdot \delta\underline{v} \, \mathrm{d}V \\
&= \int_A \left(\Big(\underline{\underline{P}} \cdot \underline{n} - \underline{\mathfrak{s}} \Big) \cdot \delta\underline{v} + \sum_i (\underline{p}_i \cdot \underline{n} - \pi_i) \, \delta\nu_i \right) \mathrm{d}A .
\end{aligned} \tag{5.13}$$

Applying the fundamental theorem of variational calculus, the local balance of forces and balance of micro forces (flow rules) for all slip systems follow as

$$\underline{\underline{P}} \cdot \underline{\nabla} + \varrho \, (\underline{f} - \ddot{\underline{u}}) = \underline{0} \, , \tag{5.14a}$$

$$\underline{p}_i \cdot \underline{\nabla} + (\tau_i - \varsigma_i) - \Lambda \frac{\partial d}{\partial \nu_i} = 0 \;\; \forall \, i \, , \tag{5.14b}$$

as well as boundary conditions

$$\underline{\underline{P}} \cdot \underline{n} = \underline{\mathfrak{s}} \quad \text{and} \quad \underline{p}_i \cdot \underline{n} = \pi_i \, \forall \, i \, . \tag{5.14c}$$

In addition, there is the symmetry condition $\underline{\underline{P}} = \underline{\underline{P}}^{\mathrm{T}}$ left over from the principle of angular momentum. This system of partial differential equations and corresponding boundary conditions (BCs) represents the *strong* formulation of equilibrium. By comparison with the Cauchy relations (3.17) it becomes clear that $\underline{\underline{P}} = \underline{\underline{\sigma}}$ is the true stress tensor and $\underline{p}_i = q_i$ is a (true) Peach-Koehler force. With this knowledge, the equilibrium conditions (5.14a) and (5.14b) correspond exactly to the field equations (3.31) and (3.32)/(3.38). Conversely, it is now also evident that the Relations (3.17) are natural BCs (5.14c). The BC $\underline{p}_i \cdot \underline{n} = \pi_i = 0$ applies to microscopically non-interactive boundaries [2, 3].

5.3 Matrix Formulation for Plane Deformation and Two Slip Systems

For programming, it is favorable to translate the weak formulation (5.8) into matrix form. For the sake of clarity, this is subsequently only presented for a plane problem with two slip systems. First, the lattice vectors of the cubic crystal have to be defined (in relation to some global $\{x, y\}$-coordinate system). It is chosen

$$\underline{\hat{a}}_{\mathrm{I}} = \cos\varphi\, \underline{e}_x + \sin\varphi\, \underline{e}_y \quad , \quad \underline{\hat{a}}_{\mathrm{II}} = -\sin\varphi\, \underline{e}_x + \cos\varphi\, \underline{e}_y \quad , \quad \underline{\hat{a}}_{\mathrm{III}} = \underline{e}_z\,, \qquad (5.15)$$

where the angle φ defines the crystal lattice rotation around the z-axis. The two slip systems $\underline{\underline{\hat{S}}}_1 = \underline{\hat{a}}_{\mathrm{II}} \otimes \underline{\hat{a}}_{\mathrm{I}}$, $\underline{\underline{\hat{S}}}_2 = \underline{\hat{a}}_{\mathrm{I}} \otimes \underline{\hat{a}}_{\mathrm{II}}$ and two displacement components lead to

$$\underline{\underline{\hat{L}}}_{\mathrm{p}} = \nu_1(x, y)\, \underline{\underline{\hat{S}}}_1 + \nu_2(x, y)\, \underline{\underline{\hat{S}}}_2 \quad \text{and} \quad \underline{u} = u_x(x, y)\, \underline{e}_x + u_y(x, y)\, \underline{e}_y\,. \qquad (5.16)$$

The coefficients of the slip rates and velocities as well as the Peach-Koehler forces $\underline{q}_i$ and the Cauchy stress $\underline{\underline{\sigma}}$ are arranged in column matrices[2]:

$$[\nu] = \begin{bmatrix} \nu_1 \\ \nu_2 \end{bmatrix}, [v] = \begin{bmatrix} v_x \\ v_y \end{bmatrix} \quad \text{and} \quad [q] = \begin{bmatrix} q_{1x} \\ q_{1y} \\ q_{2x} \\ q_{2y} \end{bmatrix}, [\sigma] = \begin{bmatrix} \sigma_{xx} \\ \sigma_{yy} \\ \sigma_{xy} \end{bmatrix}. \qquad (5.17)$$

Introducing suitable differential matrices, which contain the Cartesian coefficients of the Nabla operator, the gradients of the primary fields are represented as follows:

[2]In the following it is consistently made use of the fact that in 2D the coefficients of first-order tensors can be arranged in 2×1 matrices, coefficients of symmetric second-order tensors in 3×1 matrices and coefficients of general second-order tensors in 2×2 matrices.

$$\begin{bmatrix} \delta\nu_{1,x} \\ \delta\nu_{1,y} \\ \delta\nu_{2,x} \\ \delta\nu_{2,y} \end{bmatrix} = \underbrace{\begin{bmatrix} \partial_x & 0 \\ \partial_y & 0 \\ 0 & \partial_x \\ 0 & \partial_y \end{bmatrix}}_{[D_\nu]^\mathrm{T}} \begin{bmatrix} \delta\nu_1 \\ \delta\nu_2 \end{bmatrix} \quad \text{and} \quad \begin{bmatrix} \delta v_{x,x} \\ \delta v_{y,y} \\ \delta v_{x,y} + \delta v_{y,x} \end{bmatrix} = \underbrace{\begin{bmatrix} \partial_x & 0 \\ 0 & \partial_y \\ \partial_x & \partial_y \end{bmatrix}}_{[D_\mathrm{u}]^\mathrm{T}} \begin{bmatrix} \delta v_x \\ \delta v_y \end{bmatrix} . \tag{5.18}$$

The scalar products of the Peach-Koehler forces and virtual slip rate gradients as well as of the stress tensor and virtual velocity gradient in matrix form are thus

$$\begin{aligned} \sum_{i=1}^{2} \underline{q}_i \cdot (\underline{\nabla}\, \delta\nu_i) &\overset{2S}{=} q_{1x}\, \delta\nu_{1,x} + q_{1y}\, \delta\nu_{1,y} + q_{2x}\, \delta\nu_{2,x} + q_{2y}\, \delta\nu_{2,y} \\ &= \left(\begin{bmatrix} \partial_x & 0 \\ \partial_y & 0 \\ 0 & \partial_x \\ 0 & \partial_y \end{bmatrix} \begin{bmatrix} \delta\nu_1 \\ \delta\nu_2 \end{bmatrix} \right)^\mathrm{T} \begin{bmatrix} q_{1x} \\ q_{1y} \\ q_{2x} \\ q_{2y} \end{bmatrix} , \\ \underline{\underline{\sigma}} \cdot\cdot (\underline{\nabla} \otimes \delta\underline{v}) &\overset{2D}{=} \sigma_{xx}\, \delta v_{x,x} + \sigma_{yy}\, \delta v_{y,y} + \sigma_{xy}\, (\delta v_{x,y} + \delta v_{y,x}) \\ &= \left(\begin{bmatrix} \partial_x & 0 \\ 0 & \partial_y \\ \partial_x & \partial_y \end{bmatrix} \begin{bmatrix} \delta v_x \\ \delta v_y \end{bmatrix} \right)^\mathrm{T} \begin{bmatrix} \sigma_{xx} \\ \sigma_{yy} \\ \sigma_{xy} \end{bmatrix} . \end{aligned}$$

The scalar products of the resolved shear stresses and the associated virtual slip rates give

$$\begin{aligned} \sum_{i=1}^{2} \tau_i\, \delta\nu_i &= \sum_{i=1}^{2} \underline{\underline{\sigma}} \cdot\cdot (\underline{s}_i \otimes \underline{m}_i)^\mathrm{T} \delta\nu_i \overset{2D}{=} \sum_{i=1}^{2} \delta\nu_i \left(s_{ix}\, \sigma_{xx}\, m_{ix} + s_{iy}\, \sigma_{yy}\, m_{iy} + \sigma_{xy} (s_{ix} m_{iy} \right. \\ &\left. + s_{iy} m_{ix}) \right) = \begin{bmatrix} \delta\nu_1 \delta\nu_2 \end{bmatrix} \underbrace{\begin{bmatrix} s_{1x} m_{1x} & s_{1y} m_{1y} & (s_{1x} m_{1y} + s_{1y} m_{1x}) \\ s_{2x} m_{2x} & s_{2y} m_{2y} & (s_{2x} m_{2y} + s_{2y} m_{2x}) \end{bmatrix}}_{[S]} \begin{bmatrix} \sigma_{xx} \\ \sigma_{yy} \\ \sigma_{xy} \end{bmatrix} . \end{aligned}$$

Energetical and dissipative back stresses and boundary loads of both slip systems

$$[\varsigma] = \begin{bmatrix} \varsigma_1 \\ \varsigma_2 \end{bmatrix} \quad , \quad [\kappa] = \begin{bmatrix} \kappa_1 \\ \kappa_2 \end{bmatrix} \quad , \quad [\pi] = \begin{bmatrix} \pi_1 \\ \pi_2 \end{bmatrix} , \tag{5.19}$$

as well as the coefficients of acceleration, mass force and traction vector

$$[\ddot{u}] = \begin{bmatrix} \ddot{u}_x \\ \ddot{u}_y \end{bmatrix} \quad , \quad [f] = \begin{bmatrix} f_x \\ f_y \end{bmatrix} \quad , \quad [\mathfrak{s}] = \begin{bmatrix} \mathfrak{s}_x \\ \mathfrak{s}_y \end{bmatrix} , \tag{5.20}$$

have been arranged in column matrices, too. With all this, the weak formulation (5.8) for the special case of plane deformation with two active slip systems reads:

$$\int_V \left(\left([\delta\nu]^{\mathrm{T}} [D_\nu] \right) [q] - [\delta\nu]^{\mathrm{T}} \left([S][\sigma] - [\varsigma] - [\kappa] \right) \right) \mathrm{d}V$$
$$+ \int_V \left(\left([\delta v]^{\mathrm{T}} [D_{\mathrm{u}}] \right) [\sigma] - [\delta v]^{\mathrm{T}} \varrho \left([f] - [\ddot{u}] \right) \right) \mathrm{d}V = \int_A \left([\delta v]^{\mathrm{T}} [\mathfrak{z}] + [\delta\nu]^{\mathrm{T}} [\pi] \right) \mathrm{d}A \,. \tag{5.21}$$

For completeness, the coefficients of the matrices $[\sigma]$, $[\varsigma]$, $[\kappa]$ and $[q]$ are needed. Therefore the constitutive relations have to be evaluated, which is done firstly in the lattice space.[3] To determine the coefficients of the stress tensor, the lattice vectors (5.15) are inserted into the elasticity law (3.9). After lifting out the base and converting it to matrix notation remains

$$\begin{bmatrix} \widehat{T}^{\mathrm{e}}_{xx} \\ \widehat{T}^{\mathrm{e}}_{yy} \\ \widehat{T}^{\mathrm{e}}_{xy} \end{bmatrix} = \underbrace{[D] \begin{bmatrix} \varkappa & \lambda & 0 \\ \lambda & \varkappa & 0 \\ 0 & 0 & \mu \end{bmatrix} [D]^{\mathrm{T}}}_{[\widehat{K}]} \begin{bmatrix} \hat{E}^{\mathrm{e}}_{xx} \\ \hat{E}^{\mathrm{e}}_{yy} \\ 2\hat{E}^{\mathrm{e}}_{xy} \end{bmatrix} \quad \text{with } [D] = \begin{bmatrix} \cos^2(\varphi) & \sin^2(\varphi) & -\sin(2\varphi) \\ \sin^2(\varphi) & \cos^2(\varphi) & \sin(2\varphi) \\ \frac{1}{2}\sin(2\varphi) & -\frac{1}{2}\sin(2\varphi) & \cos(2\varphi) \end{bmatrix} . \tag{5.22}$$

The transformation matrix $[D]$ performs the rotation from crystal coordinates $\{\mathrm{I}, \mathrm{II}, \mathrm{III}\}$ to the global coordinates $\{x, y, z\}$. For a single crystal, $[D]$ and thus also the elasticity matrix $[\widehat{K}]$ are constant *in the lattice space*. The Cauchy stresses are finally obtained by means of the Relation (3.29), which takes the 2D matrix form

$$[\sigma] \equiv \begin{bmatrix} \sigma_{xx} \\ \sigma_{yy} \\ \sigma_{xy} \end{bmatrix} = \frac{1}{F^{\mathrm{e}}_{xx} F^{\mathrm{e}}_{yy} - F^{\mathrm{e}}_{xy} F^{\mathrm{e}}_{yx}} \begin{bmatrix} F^{\mathrm{e}}_{xx} F^{\mathrm{e}}_{xx} & F^{\mathrm{e}}_{xy} F^{\mathrm{e}}_{xy} & F^{\mathrm{e}}_{xx} F^{\mathrm{e}}_{xy} + F^{\mathrm{e}}_{xx} F^{\mathrm{e}}_{xy} \\ F^{\mathrm{e}}_{yx} F^{\mathrm{e}}_{yx} & F^{\mathrm{e}}_{yy} F^{\mathrm{e}}_{yy} & F^{\mathrm{e}}_{yy} F^{\mathrm{e}}_{yx} + F^{\mathrm{e}}_{yy} F^{\mathrm{e}}_{yx} \\ F^{\mathrm{e}}_{yx} F^{\mathrm{e}}_{xx} & F^{\mathrm{e}}_{xy} F^{\mathrm{e}}_{yy} & F^{\mathrm{e}}_{xx} F^{\mathrm{e}}_{yy} - F^{\mathrm{e}}_{xy} F^{\mathrm{e}}_{yx} \end{bmatrix} \begin{bmatrix} \widehat{T}^{\mathrm{e}}_{xx} \\ \widehat{T}^{\mathrm{e}}_{yy} \\ \widehat{T}^{\mathrm{e}}_{xy} \end{bmatrix} . \tag{5.23}$$

Here $j = I_3(\underline{\underline{F}}_{\mathrm{e}})$ was utilized as well as that for plane elastic/plastic deformation

$$\left[F^{\mathrm{e}}\right] = \begin{bmatrix} F^{\mathrm{e}}_{xx} & F^{\mathrm{e}}_{xy} & 0 \\ F^{\mathrm{e}}_{yx} & F^{\mathrm{e}}_{yy} & 0 \\ 0 & 0 & 1 \end{bmatrix} \quad \text{and} \quad \left[F^{\mathrm{p}}\right] = \begin{bmatrix} F^{\mathrm{p}}_{xx} & F^{\mathrm{p}}_{xy} & 0 \\ F^{\mathrm{p}}_{yx} & F^{\mathrm{p}}_{yy} & 0 \\ 0 & 0 & 1 \end{bmatrix}$$

hold. In this case the dislocation density tensor (2.18) has only two components:

$$\underline{\underline{\hat{\alpha}}}(x, y) = \underbrace{\left(\hat{\alpha}_{xz}\, \underline{e}_x + \hat{\alpha}_{yz}\, \underline{e}_y \right)}_{\underline{\hat{\alpha}}_z} \otimes \underline{e}_z \quad \text{with} \quad \begin{bmatrix} \hat{\alpha}_{xz} \\ \hat{\alpha}_{yz} \end{bmatrix} = \begin{bmatrix} F^{\mathrm{p}}_{xy,\tilde{x}} - F^{\mathrm{p}}_{xx,\tilde{y}} \\ F^{\mathrm{p}}_{yy,\tilde{x}} - F^{\mathrm{p}}_{yx,\tilde{y}} \end{bmatrix} , \tag{5.24}$$

[3] Because of the undisturbed, ideal crystal lattice the relations there are much more simple compared to a direct evaluation of the constitutive relations in the current configuration.

and contains only information from the dislocation density *vector* $\hat{\underset{\frown}{\alpha}}_z$.[4] According to Eq. (3.15), the higher order stress tensor also consists only of one dyadic product:

$$\underline{\underline{\hat{X}}}(x,y) = \underbrace{\left(\hat{X}_{xz}\,\underline{e}_x + \hat{X}_{yz}\,\underline{e}_y\right)}_{\underline{\hat{x}}_z} \otimes \underline{e}_z \quad \text{with} \quad \underline{\hat{x}}_z = \ell\left(-\frac{\partial\hat{\phi}_\mathrm{p}}{\partial r_1}\,\underline{\hat{a}}_{\mathrm{II}} + \frac{\partial\hat{\phi}_\mathrm{p}}{\partial r_2}\,\underline{\hat{a}}_{\mathrm{I}}\right). \tag{5.25}$$

For the energetical back stresses $(3.27)_1$ in the slip systems 1 and 2 it follows

$$\left(\underline{\underline{\hat{X}}}\cdot\underline{\underline{\hat{\alpha}}}^\mathrm{T} + \underline{\underline{\hat{X}}}^\mathrm{T}\cdot\underline{\underline{\hat{\alpha}}}\right)\cdot\cdot\,\underline{\underline{\hat{S}}}_i^\mathrm{T} = \underline{\hat{s}}_i\cdot\left(\underline{\hat{x}}_z\otimes\underline{\hat{\alpha}}_z + (\underline{\hat{x}}_z\cdot\underline{\hat{\alpha}}_z)\,\underline{e}_z\otimes\underline{e}_z\right)\cdot\underline{\widehat{m}}_i\,, \tag{5.26}$$

where the second summand was omitted because of $\underline{e}_z\cdot\underline{\widehat{m}}_1 = \underline{e}_z\cdot\underline{\widehat{m}}_2 = 0$. Taking into account the definition (2.26) of the dislocation densities and the internal length scale $\ell = 1/(b\rho_\mathrm{s})$, one obtains

$$\begin{bmatrix} j\,\varsigma_1 = (\underline{\hat{a}}_{\mathrm{II}}\cdot\underline{\hat{x}}_z)(\underline{\hat{\alpha}}_z\cdot\underline{\hat{a}}_{\mathrm{I}}) \\ j\,\varsigma_2 = (\underline{\hat{a}}_{\mathrm{I}}\cdot\underline{\hat{x}}_z)(\underline{\hat{\alpha}}_z\cdot\underline{\hat{a}}_{\mathrm{II}}) \end{bmatrix} \quad\rightarrow\quad [\varsigma] = \begin{bmatrix}\varsigma_1\\ \varsigma_2\end{bmatrix} = -\frac{1}{j}\begin{bmatrix}\partial\hat{\phi}_\mathrm{p}/\partial r_1\cdot r_2\\ \partial\hat{\phi}_\mathrm{p}/\partial r_2\cdot r_1\end{bmatrix}. \tag{5.27}$$

With the two proposed forms of potential $\hat{\phi}_\mathrm{p}$, compact expressions finally result as

$$\begin{bmatrix}\varsigma_1\\ \varsigma_2\end{bmatrix} \overset{(3.12)}{=} -\frac{1}{j}\,k\mu r_1 r_2\begin{bmatrix}1\\1\end{bmatrix}, \quad \begin{bmatrix}\varsigma_1\\ \varsigma_2\end{bmatrix} \overset{(3.13)}{=} -\frac{1}{j}\,k\mu r_1 r_2\begin{bmatrix}1/(1-r_1^2)\\ 1/(1-r_2^2)\end{bmatrix}, \tag{5.28}$$

which are equal for low dislocation densities $\rho_i \ll \rho_\mathrm{s}$ i.e. $r_i \ll 1$. At this point it also becomes clear that the energetic back stresses $[\varsigma]$ capture slip system interactions, since the products of *different* dislocation densities occur (cf. [4]). In addition, the Peach-Koehler forces in the lattice space read

$$\underline{\hat{q}}_i = \underline{\widehat{m}}_i\times(\underline{\hat{s}}_i\cdot\underline{\underline{\hat{X}}}) = -\underline{\hat{s}}_i\cdot\underline{\underline{\hat{X}}}\times\underline{\widehat{m}}_i = (\underline{\hat{s}}_i\cdot\underline{\hat{x}}_z)(\underline{\widehat{m}}_i\times\underline{e}_z)\,. \tag{5.29}$$

Applying Eq. (5.25) as well as the relations $\underline{\widehat{m}}_1\times\underline{e}_z = -\underline{\hat{s}}_1$ and $\underline{\widehat{m}}_2\times\underline{e}_z = \underline{\hat{s}}_2$ yields

$$\underline{\hat{q}}_1 = -(\underline{\hat{a}}_{\mathrm{II}}\cdot\underline{\hat{x}}_z)\,\underline{\hat{s}}_1 \qquad = (\partial\hat{\phi}_\mathrm{p}/\partial r_1)\,\underline{\hat{s}}_1, \tag{5.30}$$

$$\underline{\hat{q}}_2 = (\underline{\hat{a}}_{\mathrm{I}}\cdot\underline{\hat{x}}_z)\,\underline{\hat{s}}_2 \qquad = (\partial\hat{\phi}_\mathrm{p}/\partial r_2)\,\underline{\hat{s}}_2. \tag{5.31}$$

The true Peach-Koehler forces $\underline{q}_i = \frac{1}{j}\underline{\underline{F}}_\mathrm{e}\cdot\underline{\hat{q}}_i$ according to Definition $(3.27)_2$ finally read

---

[4] Multiplying $\mathrm{Curl}(\underline{\underline{F}}_\mathrm{p})$ with $\underline{\underline{F}}_\mathrm{p}^\mathrm{T}$ has no effect for plane deformation since $\underline{\underline{F}}_\mathrm{p}\cdot\underline{e}_z = \underline{e}_z$ holds.

$$\underline{q}_1 = \frac{1}{j}(\partial\hat{\phi}_p/\partial r_1)\,\underline{s}_1 \overset{(3.12)}{=} \frac{1}{j}\,k\mu\ell\, r_1\,\underline{a}_{\mathrm{II}} = q_{1x}\,\underline{e}_x + q_{1y}\,\underline{e}_y\,, \tag{5.32a}$$

$$\underline{q}_2 = \frac{1}{j}(\partial\hat{\phi}_p/\partial r_2)\,\underline{s}_2 \overset{(3.12)}{=} \frac{1}{j}\,k\mu\ell\, r_2\,\underline{a}_{\mathrm{I}} = q_{2x}\,\underline{e}_x + q_{2y}\,\underline{e}_y\,, \tag{5.32b}$$

where the lattice vectors were transformed into the current configuration according to Rule (2.3). Thus, the Peach-Koehler forces of the edge dislocations act along the current slip directions.[5] Finally the dissipative shear stresses κ_i are given, which are derived from the dissipation potential (3.44) with $\Lambda = 1/n = \frac{p}{1+p}$ to

$$\begin{bmatrix}\kappa_1\\ \kappa_2\end{bmatrix} = \Lambda \begin{bmatrix}\partial d/\partial\nu_1\\ \partial d/\partial\nu_2\end{bmatrix} = \tau_{\mathrm{cr}} \begin{bmatrix}(\nu_1/\bar{\nu})^{1/p}\\ (\nu_2/\bar{\nu})^{1/p}\end{bmatrix}\,, \tag{5.33}$$

where the constant reference value $\bar{\nu} = \frac{\tau_{\mathrm{cr}}}{\eta_{\mathrm{p}}}$ represents a characteristic slip rate.

References

1. Pao, Y.H., Wang, L.S., Chen, K.C.: Principle of virtual power for thermomechanics of fluids and solids with dissipation. Int. J. Eng. Sci. **49**(12), 1502–1516 (2011)
2. Gurtin, M.E.: A finite-deformation, gradient theory of single-crystal plasticity with free energy dependent on densities of geometrically necessary dislocations. Int. J. Plast. **24**(4), 702–725 (2008)
3. Gurtin, M.E.: A finite-deformation, gradient theory of single-crystal plasticity with free energy dependent on the accumulation of geometrically necessary dislocations. Int. J. Plast. **26**(8), 1073–1096 (2010)
4. Gurtin, M.E.: A gradient theory of single-crystal viscoplasticity that accounts for geometrically necessary dislocations. J. Mech. Phys. Solids **50**(1), 5–32 (2002)

[5]For screw dislocations, the Peach-Koehler forces would act in the same plane but perpendicular to the slip directions.

Chapter 6
Numerical Solution with the Finite Element Method

Abstract This chapter presents a comprehensive methodology for the numerical solution of the variational formulation of the geometrically nonlinear continuum dislocation theory by means of the finite element method. For the special case of a continuously dislocated, plane, cubic primitive single crystal with two active slip systems, the discretized integral equations for one finite element are transformed to matrix notation.

6.1 Weak Form of Equilibrium for One Finite Element

The initial and boundary value problem (5.14) does not allow an analytical solution in general. For this reason, the finite element method is used. A comprehensive overview of this topic in the context of crystal plasticity already exists [1]. The starting point for the finite element method is the weak formulation (5.8). The basic domain Ω is split into a number of N_e simple subdomains Ω_e (elements), so that $\Omega = \cup\, \Omega_e$ is. The displacements $\underline{u}$ and the slip rates $\boldsymbol{\nu}$ are regarded as independent, primary variables. This choice has two advantages: (i) the set of sought field quantities is minimal and (ii) for both types of field quantities boundary conditions of the first and second kind can be directly established *and* interpreted. Typical for FEM [2], the shape of the elements is interpolated with the shape functions $G_L(\underline{r})$, where the grid points $\underline{r}_L$ are called nodes. Here applies

$$\underline{r}(t) = \sum_L G_L(\underline{r})\, \underline{r}_L(t) \quad \text{with} \quad G_L(\underline{r}_M) = \delta_{LM} \quad \text{and} \quad \sum_L G_L = 1\,. \qquad (6.1)$$

According to the isoparametric concept, the primary fields are interpolated with the same shape functions G_L so that the displacements and slip rates in the element are

C. B. Silbermann et al., *Introduction to Geometrically Nonlinear Continuum Dislocation Theory*, SpringerBriefs in Continuum Mechanics,
https://doi.org/10.1007/978-3-030-63696-8_6

$$\underline{u}(\underline{r},t)=\sum_L G_L(\underline{r})\,\underline{u}_L(t)\quad,\quad \nu_i(\underline{r},t)=\sum_L G_L(\underline{r})\,\nu_{iL}(t)\,. \tag{6.2}$$

The unknown nodal values $\underline{u}_L(t)$ and $\nu_{iL}(t)$ depend on the time. For the actual velocity and acceleration fields, this results directly in

$$\underline{v}(\underline{r},t)=\dot{\underline{u}}(\underline{r},t)=\sum_L G_L(\underline{r})\,\dot{\underline{u}}_L(t)\quad,\quad \ddot{\underline{u}}(\underline{r},t)=\sum_L G_L(\underline{r})\,\ddot{\underline{u}}_L(t)\,. \tag{6.3}$$

According to the Bubnov-Galerkin method (cf. e. g. [2]), the same approach $(6.2)_2$ / $(6.3)_1$ is chosen for the virtual slip rate and velocity fields as for the actual fields:

$$\delta\underline{v}(\underline{r},t)=\sum_L G_L(\underline{r})\,\delta\underline{v}_L(t)\quad,\quad \delta\nu_i(\underline{r},t)=\sum_L G_L(\underline{r})\,\delta\nu_{iL}(t) \tag{6.4}$$

with the virtual nodal values $\delta\underline{v}_L$ and $\delta\nu_{iL}$. Finally, the virtual gradients read

$$\underline{\nabla}\otimes\delta\underline{v}(\underline{r},t)=\sum_L(\underline{\nabla}G_L(\underline{r}))\otimes\delta\underline{v}_L(t)\quad,\quad \underline{\nabla}\delta\nu_i(\underline{r},t)=\sum_L(\underline{\nabla}G_L(\underline{r}))\,\delta\nu_{iL}(t). \tag{6.5}$$

Thus the weak form of equilibrium *for one finite element* now reads

$$\sum_i\left\{\sum_L\int_V\Big(\underline{q}_i\cdot(\underline{\nabla}G_L)-(\tau_i-\varsigma_i-\kappa_i)\,G_L\Big)\,\mathrm{d}V-\sum_L\int_A\pi_i\,G_L\,\mathrm{d}A\right\}\delta\nu_{iL}$$
$$+\left\{\sum_L\int_V\Big(\underline{\underline{\sigma}}\cdot(\underline{\nabla}G_L)-\varrho\,(\underline{f}-\ddot{\underline{u}})\,G_L\Big)\,\mathrm{d}V-\sum_L\int_A\underline{s}\,G_L\,\mathrm{d}A\right\}\cdot\delta\underline{v}_L=0\,, \tag{6.6}$$

whereby the virtual nodal values could be excluded due to their spatial constancy. The expressions between the curly brackets correspond to force variables that are assigned to the virtual nodal values. From the fundamental theorem of variational calculus it follows for independent and arbitrary $\delta\underline{v}_L$, $\delta\nu_{iL}$ that these expressions must disappear—which equals a weak formulation of the equilibrium of forces:

$$\underbrace{\sum_L\int_V\Big(\underline{q}_i\cdot(\underline{\nabla}G_L)-\big(\underline{s}_i\cdot\underline{\underline{\sigma}}\cdot\underline{m}_i-\varsigma_i-\kappa_i\big)\,G_L\Big)\,\mathrm{d}V}_{Q_{iL}^{\mathrm{in}}}=\underbrace{\sum_L\int_A\pi_i\,G_L\,\mathrm{d}A}_{Q_{iL}^{\mathrm{ex}}}\quad\forall\,i,$$

$$\underbrace{\sum_{L,M}\int_V\varrho G_L G_M\mathrm{d}V\,\ddot{\underline{u}}_M}_{M_{LM}}+\underbrace{\sum_L\int_V\underline{\underline{\sigma}}\cdot(\underline{\nabla}G_L)\mathrm{d}V}_{\underline{F}_L^{\mathrm{in}}}=\underbrace{\sum_L\left(\int_A\underline{s}G_L\mathrm{d}A+\int_V\varrho\underline{f}\,G_L\mathrm{d}V\right)}_{\underline{F}_L^{\mathrm{ex}}}. \tag{6.7}$$

The bilinear form M_{LM} constitutes the mass matrix. In contrast to the strong formulation (5.14) in form of differential equations the equilibrium conditions in the weak FE formulation, i. e.

$$\sum_{L,M} M_{LM}\,\ddot{\underline{u}}_M + \sum_L \underline{F}_L^{\text{in}} = \sum_L \underline{F}_L^{\text{ex}} \quad , \quad \sum_L Q_{iL}^{\text{in}} = \sum_L Q_{iL}^{\text{ex}} \quad \forall\, i \tag{6.8}$$

are not fulfilled at every point of the element, but only *in the weighted mean*. The integral equations (6.7) are a strongly coupled problem, because the inner nodal forces Q_{iL}^{in} and $\underline{F}_L^{\text{in}}$ depend on *all* unknown nodal values. The coupling originates from (geometrically nonlinear) kinematics and is expressed kinetically in the form of resolved shear stresses $\tau_i = \underline{s}_i \cdot \underline{\underline{\sigma}} \cdot \underline{m}_i$. Consequently, the equation system should be solved monolithically, which is done in the following section for plane deformations and two orthogonal slip systems.

6.2 FE Solution Algorithm for Plane Deformations and Two Slip Systems

In order to obtain an FEM equation system in matrix form for plane deformation with two slip systems, the FE formulation of the force equilibrium (6.7) and the matrix representation (5.21) of the principle of virtual power are combined. For a monolithic treatment, the unknowns are arranged to the generalized tuple

$$[w] := [v_x(x,y),\ v_y(x,y),\ \nu_1(x,y),\ \nu_2(x,y)]^{\text{T}}.$$

The velocity field and the slip rate field result from this by means of

$$\begin{bmatrix} v_x \\ v_y \end{bmatrix} = \begin{bmatrix} 1 & 0 & 0 & 0 \\ 0 & 1 & 0 & 0 \end{bmatrix} \begin{bmatrix} v_x \\ v_y \\ \nu_1 \\ \nu_2 \end{bmatrix} \quad \text{and} \quad \begin{bmatrix} \nu_1 \\ \nu_2 \end{bmatrix} = \begin{bmatrix} 0 & 0 & 1 & 0 \\ 0 & 0 & 0 & 1 \end{bmatrix} \begin{bmatrix} v_x \\ v_y \\ \nu_1 \\ \nu_2 \end{bmatrix} . \tag{6.9}$$

With the unity matrix $[\text{I}] = \left[\begin{smallmatrix} 1 & 0 \\ 0 & 1 \end{smallmatrix}\right]$ and the zero matrix $[0] = \left[\begin{smallmatrix} 0 & 0 \\ 0 & 0 \end{smallmatrix}\right]$, this is written as

$$[v] = [\,\text{I}\,|\,0\,][w] \quad \text{and} \quad [\nu] = [\,0\,|\,\text{I}\,][w]\,. \tag{6.10}$$

Next, all nodal free values of the element are arranged in a column matrix

$$[\overset{e}{w}] = \left[\overset{1}{v}_x, \overset{1}{v}_y, \overset{1}{\nu}_1, \overset{1}{\nu}_2\ ,\ \overset{2}{v}_x, \overset{2}{v}_y, \overset{2}{\nu}_1, \overset{2}{\nu}_2\ ,\ \ldots \overset{\text{N}}{v}_x, \overset{\text{N}}{v}_y, \overset{\text{N}}{\nu}_1, \overset{\text{N}}{\nu}_2, \right]^{\text{T}} . \tag{6.11}$$

Interpolation (6.2) now reads in matrix form $[w] = [G]^{\text{T}}[\overset{e}{w}]$ with the matrix of shape functions

$$[G]^{\mathrm{T}} = \begin{bmatrix} G_1 & 0 & 0 & 0 & G_2 & 0 & 0 & 0 & \dots & G_N & 0 & 0 & 0 \\ 0 & G_1 & 0 & 0 & 0 & G_2 & 0 & 0 & \dots & 0 & G_N & 0 & 0 \\ 0 & 0 & G_1 & 0 & 0 & 0 & G_2 & 0 & \dots & 0 & 0 & G_N & 0 \\ 0 & 0 & 0 & G_1 & 0 & 0 & 0 & G_2 & \dots & 0 & 0 & 0 & G_N \end{bmatrix}, \tag{6.12}$$

where $G_L = G_L(x, y)$ holds for all shape functions. The element's displacements and slip rates follow thus as $[v] = [G_{\mathrm{u}}]^{\mathrm{T}}[\overset{e}{w}]$ and $[\nu] = [G_{\nu}]^{\mathrm{T}}[\overset{e}{w}]$ with

$$[G_u]^{\mathrm{T}} = [\,\mathrm{I}\,|\,0\,]\,[G]^{\mathrm{T}} = \begin{bmatrix} G_1 & 0 & 0 & 0 & G_2 & 0 & 0 & 0 & \dots & G_N & 0 & 0 & 0 \\ 0 & G_1 & 0 & 0 & 0 & G_2 & 0 & 0 & \dots & 0 & G_N & 0 & 0 \end{bmatrix}, \tag{6.13}$$

$$[G_\nu]^{\mathrm{T}} = [\,0\,|\,\mathrm{I}\,]\,[G]^{\mathrm{T}} = \begin{bmatrix} 0 & 0 & G_1 & 0 & 0 & 0 & G_2 & 0 & \dots & 0 & 0 & G_N & 0 \\ 0 & 0 & 0 & G_1 & 0 & 0 & 0 & G_2 & \dots & 0 & 0 & 0 & G_N \end{bmatrix}. \tag{6.14}$$

Inserting $[\delta v]^{\mathrm{T}} = [\delta \overset{e}{w}]^{\mathrm{T}}[G_{\mathrm{u}}]$ and $[\delta \nu]^{\mathrm{T}} = [\delta \overset{e}{w}]^{\mathrm{T}}[G_{\nu}]$ in the matrix representation (5.21) and factoring out the virtual nodal values $[\delta \overset{e}{w}]^{\mathrm{T}}$ yields

$$\begin{aligned} &[\delta \overset{e}{w}]^{\mathrm{T}} \int\limits_V \Big(\big([G_\nu][D_\nu] \big)[q] - [G_\nu]\big([S][\sigma] - [\varsigma] - [\kappa] \big) \Big) \mathrm{d}V + [\delta \overset{e}{w}]^{\mathrm{T}} \int\limits_V \Big(\big([G_{\mathrm{u}}][D_{\mathrm{u}}] \big)[\sigma] \\ &\quad - [G_{\mathrm{u}}]\big(\varrho[f] - \varrho[\ddot{u}] \big) \Big) \mathrm{d}V = [\delta \overset{e}{w}]^{\mathrm{T}} \int\limits_A \Big([G_{\mathrm{u}}][\mathfrak{s}] + [G_\nu][\pi] \Big)\, \mathrm{d}A\,. \end{aligned} \tag{6.15}$$

Inserting the accelerations in the form $[\ddot{u}] = [G_{\mathrm{u}}]^{\mathrm{T}}[\overset{e}{\dot{w}}]$ and applying the fundamental theorem of variational calculus, the following system of nonlinear equations results:

$$\begin{aligned} &\int\limits_V \varrho\,[G_{\mathrm{u}}][G_{\mathrm{u}}]^{\mathrm{T}} \mathrm{d}V\,[\overset{e}{\dot{w}}] + \int\limits_V \Big(\big([G_{\mathrm{u}}][D_{\mathrm{u}}] - [G_\nu][S] \big)[\sigma] + [G_\nu]\big([\varsigma] + [\kappa] \big) \\ &\quad + \big([G_\nu][D_\nu] \big)[q] \Big) \mathrm{d}V - \int\limits_V \varrho\,[f]\,\mathrm{d}V + \int\limits_A \Big([G_{\mathrm{u}}][\mathfrak{s}] + [G_\nu][\pi] \Big)\, \mathrm{d}A = [0]\,. \end{aligned} \tag{6.16}$$

In the following, only quasistatic processes are considered, whereby the inertia term may be neglected. Furthermore, no far fields are to act, whereby $[f]$ disappears. By introducing the abbreviations $[G_{\mathrm{u}}][D_{\mathrm{u}}] =: [B_{\mathrm{u}}]$ and $[G_\nu][D_\nu] =: [B_\nu]$ for the derivatives of the shape functions, the compact, monolithic FEM equation system

$$\int\limits_V \Big(\underbrace{\big([B_{\mathrm{u}}] - [G_\nu][S] \big)}_{(4N\times 3)} \underbrace{[\sigma]}_{3\times 1} + \underbrace{[G_\nu]}_{4N\times 2} \underbrace{\big([\varsigma] + [\kappa] \big)}_{(2\times 1)} + \underbrace{[B_\nu]}_{4N\times 4} \underbrace{[q]}_{4\times 1} \Big) \mathrm{d}V = \int\limits_A \Big(\underbrace{[G_{\mathrm{u}}][\mathfrak{s}] + [G_\nu][\pi]}_{4N\times 1} \Big) \mathrm{d}A \tag{6.17}$$

is obtained for quasistatic, plane deformation and plastic slip of a cubic primitive, continuously dislocated single crystal. This generalized force equilibrium of the form $[\overset{e}{r}_{\mathrm{in}}] = [\overset{e}{r}_{\mathrm{ex}}]$ can be transformed into the zero form [3, p. 149]

$$[\overset{e}{r}] := [\overset{e}{r}_{\mathrm{in}}] - [\overset{e}{r}_{\mathrm{ex}}] = [0]\,, \tag{6.18}$$

where $[\overset{e}{r}_{\mathrm{in}}]$ includes the inner force quantities and $[\overset{e}{r}_{\mathrm{ex}}]$ the ones impressed from outside. The column matrix $[\overset{e}{r}]$ is called residual because it vanishes in equilibrium. The solution of the equilibrium conditions thus becomes a zero search that can be solved with the Newton-Raphson method. The degrees of freedom (DOFs) are arranged analogous to the column matrix (6.11) as

$$[\overset{e}{\varpi}] = \left[\overset{1}{u}_x, \overset{1}{u}_y, \overset{1}{\nu}_1, \overset{1}{\nu}_2\,,\ \overset{2}{u}_x, \overset{2}{u}_y, \overset{2}{\nu}_1, \overset{2}{\nu}_2\,,\ \ldots \overset{\mathrm{N}}{u}_x, \overset{\mathrm{N}}{u}_y, \overset{\mathrm{N}}{\nu}_1, \overset{\mathrm{N}}{\nu}_2\right]^{\mathrm{T}}, \tag{6.19}$$

however with the difference that now the displacements—the time integrals of the velocities—are regarded as searched for.[1] From the column matrix of residuals

$$[\overset{e}{r}] = \left[\overset{1}{r}_x, \overset{1}{r}_y, \overset{1}{r}_1, \overset{1}{r}_2\,,\ \overset{2}{r}_x, \overset{2}{r}_y, \overset{2}{r}_1, \overset{2}{r}_2\,,\ \ldots \overset{\mathrm{N}}{r}_x, \overset{\mathrm{N}}{r}_y, \overset{\mathrm{N}}{r}_1, \overset{\mathrm{N}}{r}_2\right]^{\mathrm{T}} \tag{6.20}$$

follows the quadratic element stiffness matrix with the partial derivatives of the residual coefficients with respect to the individual element DOFs, i. e.

$$\left[\overset{e}{K}_{LM}\right] = \left[\frac{\partial \overset{e}{r}_L}{\partial \overset{e}{\varpi}_M}\right]^{\mathrm{T}} = \begin{bmatrix} \dfrac{\partial \overset{1}{r}_x}{\partial \overset{1}{u}_x} & \dfrac{\partial \overset{1}{r}_y}{\partial \overset{1}{u}_x} & \dfrac{\partial \overset{1}{r}_1}{\partial \overset{1}{u}_x} & \dfrac{\partial \overset{1}{r}_2}{\partial \overset{1}{u}_x} & \cdots & \dfrac{\partial \overset{\mathrm{N}}{r}_2}{\partial \overset{1}{u}_x} \\ \dfrac{\partial \overset{1}{r}_x}{\partial \overset{1}{u}_y} & \dfrac{\partial \overset{1}{r}_y}{\partial \overset{1}{u}_y} & \dfrac{\partial \overset{1}{r}_1}{\partial \overset{1}{u}_y} & \dfrac{\partial \overset{1}{r}_2}{\partial \overset{1}{u}_y} & \cdots & \dfrac{\partial \overset{\mathrm{N}}{r}_2}{\partial \overset{1}{u}_y} \\ \vdots & \vdots & \vdots & \vdots & \ddots & \vdots \\ \dfrac{\partial \overset{1}{r}_x}{\partial \overset{\mathrm{N}}{\nu}_2} & \dfrac{\partial \overset{1}{r}_y}{\partial \overset{\mathrm{N}}{\nu}_2} & \dfrac{\partial \overset{1}{r}_1}{\partial \overset{\mathrm{N}}{\nu}_2} & \dfrac{\partial \overset{1}{r}_2}{\partial \overset{\mathrm{N}}{\nu}_2} & \cdots & \dfrac{\partial \overset{\mathrm{N}}{r}_2}{\partial \overset{\mathrm{N}}{\nu}_2} \end{bmatrix}^{\mathrm{T}} . \tag{6.21}$$

In the present case, the element stiffness matrix is not symmetrical, which is due to the additional DOFs in the form of slip rates. In this study $[K^{e}_{LM}]$ is approximated by a *numerical* derivative. For this, the procedure described in [3, p. 154] is used. The exact *analytical* derivation of the element stiffness matrix is very complicated for nonlinear multi-field problems, but seems worthwhile for an established model in the future.

To determine the residual, i. e. the internal and external forces, the primary fields and their spatial derivatives are needed. The location within a finite element is param-

[1] For the slip rates, this procedure is not appropriate, since the geometrically nonlinear theory is easier and more consistent to formulate, without the slip itself being introduced [4].

eterized as usual with natural element coordinates $\chi, \eta \in [-1, +1]$, so that

$$\underline{r} = \sum_L G_L(\chi, \eta)\, \underline{r}_L \quad \text{resp.} \quad \begin{bmatrix} x \\ y \end{bmatrix} = \sum_L G_L(\chi, \eta) \begin{bmatrix} x_L \\ y_L \end{bmatrix} \tag{6.22}$$

holds according to the isoparametric concept. The primary fields are then functions $\underline{u}(\chi, \eta)$ and $\nu_i(\chi, \eta)$ of the element coordinates. To calculate $\text{grad}(\underline{u})$ and $\text{grad}(\nu_i)$ in the current configuration, it is proceeded as follows [3, p. 106]: The displacement and slip rate gradients with respect to element coordinates read

$$(\underline{\breve{\nabla}} \otimes \underline{u})^{\mathrm{T}} \mathrel{\widehat{=}} \begin{bmatrix} u_{x,\chi} & u_{x,\eta} \\ u_{y,\chi} & u_{y,\eta} \end{bmatrix} \quad \text{and} \quad \underline{\breve{\nabla}} \nu_i \mathrel{\widehat{=}} \begin{bmatrix} \nu_{i,\chi} \\ \nu_{i,\eta} \end{bmatrix}^{\mathrm{T}} . \tag{6.23}$$

Next, differentiating the basic kinematic relation $\underline{r} = \underline{\tilde{r}} + \underline{u}$ w. r. t. element coordinates, a rule for the calculation of the derivatives of the current physical coordinates w. r. t. the element coordinates is obtained:

$$(\underline{\breve{\nabla}} \otimes \underline{r})^{\mathrm{T}} = (\underline{\breve{\nabla}} \otimes \underline{\tilde{r}})^{\mathrm{T}} + (\underline{\breve{\nabla}} \otimes \underline{u})^{\mathrm{T}} , \tag{6.24}$$

$$\underline{\underline{j}} = \underline{\underline{J}} + (\underline{\breve{\nabla}} \otimes \underline{u})^{\mathrm{T}} . \tag{6.25}$$

In addition to this *additive* relation there is the *multiplicative* relation $\underline{\underline{j}} \cdot \underline{\underline{J}}^{-1} = \underline{\underline{F}}$. The coefficients of $\underline{\underline{j}}$ and $\underline{\underline{J}}$ form the Jacobi matrices of two mappings:

$$(x, y) \mapsto (\chi, \eta) \; : \; [j] = \begin{bmatrix} x_{,\chi} & x_{,\eta} \\ y_{,\chi} & y_{,\eta} \end{bmatrix} \quad \text{and} \quad (\tilde{x}, \tilde{y}) \mapsto (\chi, \eta) \; : \; [J] = \begin{bmatrix} \tilde{x}_{,\chi} & \tilde{x}_{,\eta} \\ \tilde{y}_{,\chi} & \tilde{y}_{,\eta} \end{bmatrix} . \tag{6.26}$$

The determinants of the Jacobi matrices correspond to *local* surface ratios, i. e.

$$\mathrm{d}A = \mathrm{d}x\, \mathrm{d}y = \det[j]\, \mathrm{d}\chi\, \mathrm{d}\eta \quad \text{and} \quad \mathrm{d}\tilde{A} = \mathrm{d}\tilde{x}\, \mathrm{d}\tilde{y} = \det[J]\, \mathrm{d}\chi\, \mathrm{d}\eta \,, \tag{6.27}$$

with which area integrals can be transformed to the unit square $(-1, 1) \times (-1, 1)$. With $\underline{\underline{j}}$ it is now possible to transform derivatives w. r. t. the local element coordinates to derivatives w. r. t. the current global coordinates:

$$\text{grad}(\underline{u}) = (\underline{\nabla} \otimes \underline{u})^{\mathrm{T}} = (\underline{\breve{\nabla}} \otimes \underline{u})^{\mathrm{T}} \cdot \underline{\underline{j}}^{-1} \quad \text{and} \quad \text{grad}(\nu_i) = \underline{\nabla} \nu_i = (\underline{\breve{\nabla}} \nu_i) \cdot \underline{\underline{j}}^{-1} . \tag{6.28}$$

The validity of the transform relations becomes clear in matrix notation:

$$\begin{bmatrix} u_{x,x} & u_{x,y} \\ u_{y,x} & u_{y,y} \end{bmatrix} = \begin{bmatrix} u_{x,\chi} & u_{x,\eta} \\ u_{y,\chi} & u_{y,\eta} \end{bmatrix} \begin{bmatrix} \chi_{,x} & \chi_{,y} \\ \eta_{,x} & \eta_{,y} \end{bmatrix} \quad \text{and} \quad \begin{bmatrix} \nu_{i,x} & \nu_{i,y} \end{bmatrix} = \begin{bmatrix} \nu_{i,\chi} & \nu_{i,\eta} \end{bmatrix} \begin{bmatrix} \chi_{,x} & \chi_{,y} \\ \eta_{,x} & \eta_{,y} \end{bmatrix} .$$

Both $\text{grad}(\underline{u})$ and $\text{Grad}(\underline{u})$ can be used to immediately calculate the deformation

$$\underline{\underline{F}} = \text{Grad}(\underline{u}) + \underline{\underline{I}} = \left(\underline{\underline{I}} - \text{grad}(\underline{u})\right)^{-1} \quad \text{and} \quad (\underline{\widetilde{\nabla}} \otimes \underline{u})^{\text{T}} = (\underline{\nabla} \otimes \underline{u})^{\text{T}} \cdot \underline{\underline{F}} \,. \quad (6.29)$$

In order to set up the FEM equation system (6.17) at element level, the required kinetic quantities must be calculated using the known fields $\underline{u}(\underline{r})$, $\boldsymbol{\nu}(\underline{r})$ and $\text{grad}(\underline{u})$, $\text{grad}(\nu_i)$. Specifically these are the Cauchy stresses (5.23) as well as for two slip systems the distributed Peach-Koehler forces (5.32) and the energetic/dissipative back stresses (5.27) and (5.33). In order to proceed this way, a time integration of kinematic relations is necessary, which is shown in the next section.

6.3 Time Integration and History Variables

Under the assumption that $\underline{\underline{\hat{L}}}_{\text{p}}$ remains constant during some time step of width Δt, the ordinary tensor differential equation $\dot{\underline{\underline{F}}}_{\text{p}} = \underline{\underline{\hat{L}}}_{\text{p}} \cdot \underline{\underline{F}}_{\text{p}}$ has the analytical solution

$$\underline{\underline{F}}_{\text{p}}(t+\Delta t) = \exp\left(\!\!\left(\Delta t\, \underline{\underline{\hat{L}}}_{\text{p}}(t+\Delta t)\right)\!\!\right) \cdot \underline{\underline{F}}_{\text{p}}(t) =: \underline{\underline{\hat{e}}}(t+\Delta t) \cdot \underline{\underline{F}}_{\text{p}}(t) \qquad (6.30)$$

for $\underline{\underline{F}}_{\text{p}}$ at the new time instant $t+\Delta t$ (cf. e. g. [5, p. 9]). In the context of nonlinear continuum dilsocation theory and slip system approach (2.7) Formula (2.12) follows, i.e.

$$\underline{\underline{\hat{e}}}(t+\Delta t) = \exp\left(\!\!\left(\Delta t\, \underline{\underline{\hat{L}}}_{\text{p}}(t+\Delta t)\right)\!\!\right) = \exp\left(\!\!\left(\sum_i \nu_i(t+\Delta t)\, \Delta t\, \underline{\hat{s}}_i \otimes \underline{\widehat{m}}_i\right)\!\!\right) . \qquad (6.31)$$

The tensor exponential function $\exp((\cdot))$ can be determined from a power series (cf. e. g. [5, p. 9]). In practice, in most cases this is only possible approximately. For the special case of plane deformation with two orthogonal slip systems, i. e.

$$\underline{\underline{F}}_{\text{p}}(t+\Delta t) = \exp\left(\left(\xi_1\, \underline{\hat{a}}_{\text{II}} \otimes \underline{\hat{a}}_{\text{I}} + \xi_2\, \underline{\hat{a}}_{\text{I}} \otimes \underline{\hat{a}}_{\text{II}}\right)\right) \cdot \underline{\underline{F}}_{\text{p}}(t) \ \text{ with } \ \xi_i := \nu_i(t+\Delta t)\, \Delta t \,, \qquad (6.32)$$

there is, however, an exact analytical solution of the tensor exponential (cf. Appendix A.2.1), which was also used for the FE implementation. With known $\underline{\underline{F}}_{\text{p}}(t+\Delta t)$ from solution (6.30) the current dislocation density tensor can be calculated as follows:

$$\begin{aligned}
\underline{\underline{\hat{\alpha}}}(t+\Delta t) &= \text{Curl}\left(\underline{\underline{F}}_{\text{p}}(t+\Delta t)\right) \cdot \underline{\underline{F}}_{\text{p}}^{\text{T}}(t+\Delta t) = \left(-\underline{\underline{F}}_{\text{p}}(t+\Delta t) \times \underline{\widetilde{\nabla}}\right) \cdot \underline{\underline{F}}_{\text{p}}^{\text{T}}(t+\Delta t) \\
&= \left\{\underline{\underline{\hat{e}}} \cdot \left(-\underline{\underline{F}}_{\text{p}}(t) \times \underline{\widetilde{\nabla}}\right) \cdot \underline{\underline{F}}_{\text{p}}^{\text{T}}(t) + \left(\widetilde{\nabla}_a\, \underline{\underline{\hat{e}}}\right) \cdot \left(-\, \underline{\underline{F}}_{\text{p}}(t) \times \underline{e}_a\right) \cdot \underline{\underline{F}}_{\text{p}}^{\text{T}}(t)\right\} \cdot \underline{\underline{\hat{e}}}^{\text{T}} .
\end{aligned} \qquad (6.33)$$

Because of the special form (6.31) the derivative of the tensor exponential $\underline{\underline{\hat{e}}}$ reads

$$\widetilde{\nabla}_a \underline{\underline{\hat{e}}} = \sum_i \frac{\partial \underline{\underline{\hat{e}}}(t+\Delta t)}{\partial \nu_i(t+\Delta t)} \frac{\partial \nu_i(t+\Delta t)}{\partial \tilde{x}_a} =: \sum_i \underline{\underline{\hat{e}}}_{,i} \left(\widetilde{\nabla}_a \, \nu_i(t+\Delta t) \right) . \tag{6.34}$$

With $(-\underline{\underline{F}}_{\mathrm{p}}(t) \times \underline{\widetilde{\nabla}}) \cdot \underline{\underline{F}}_{\mathrm{p}}^{\mathrm{T}}(t) = \underline{\underline{\hat{\alpha}}}(t)$ and $(-\underline{\underline{F}}_{\mathrm{p}} \times \underline{e}_a) \cdot \underline{\underline{F}}_{\mathrm{p}}^{\mathrm{T}} = \underline{\underline{\epsilon}} \cdot \underline{\underline{F}}_{\mathrm{p}}^{-\mathrm{T}} \cdot \underline{e}_a$ it follows further (cf. Appendix A.1.1)

$$\underline{\underline{\hat{\alpha}}}(t+\Delta t) = \left\{ \underline{\underline{\hat{e}}} \cdot \underline{\underline{\hat{\alpha}}}(t) - \sum_i \underline{\underline{\hat{e}}}_{,i} \times \left(\underline{\underline{F}}_{\mathrm{p}}^{-\mathrm{T}}(t) \cdot \underline{\widetilde{\nabla}} \nu_i(t+\Delta t) \right) \right\} \cdot \underline{\underline{\hat{e}}}^{\mathrm{T}} . \tag{6.35}$$

In order to eliminate $\underline{\widetilde{\nabla}} = \underline{\underline{F}}^{\mathrm{T}} \cdot \underline{\nabla}$ the solution (6.30) is inserted in the conversion

$$\begin{aligned} \underline{\underline{F}}_{\mathrm{p}}^{-\mathrm{T}}(t) \cdot \underline{\widetilde{\nabla}} \nu_i(t+\Delta t) &= \underline{\underline{F}}_{\mathrm{p}}^{-\mathrm{T}}(t) \cdot \underline{\underline{F}}^{\mathrm{T}}(t+\Delta t) \cdot \underline{\nabla} \nu_i(t+\Delta t) \\ &= \underline{\underline{F}}_{\mathrm{p}}^{-\mathrm{T}}(t) \cdot \underline{\underline{F}}_{\mathrm{p}}^{\mathrm{T}}(t+\Delta t) \cdot \underline{\underline{F}}_{\mathrm{e}}^{\mathrm{T}}(t+\Delta t) \cdot \underline{\nabla} \nu_i(t+\Delta t), \end{aligned}$$

such that with the abbreviation $\underline{\widehat{\nabla}} \nu_i(t+\Delta t) = \underline{\underline{F}}_{\mathrm{e}}^{\mathrm{T}}(t+\Delta t) \cdot \underline{\nabla} \nu_i(t+\Delta t)$ the result is

$$\underline{\underline{\hat{\alpha}}}(t+\Delta t) = \left\{ \underline{\underline{\hat{e}}}(t+\Delta t) \cdot \underline{\underline{\hat{\alpha}}}(t) - \sum_i \underline{\underline{\hat{e}}}_{,i}(t+\Delta t) \times \left(\underline{\underline{\hat{e}}}^{\mathrm{T}}(t+\Delta t) \cdot \underline{\widehat{\nabla}} \nu_i(t+\Delta t) \right) \right\} \cdot \underline{\underline{\hat{e}}}^{\mathrm{T}}(t+\Delta t) . \tag{6.36}$$

This formula can be interpreted as a solution of the differential equation (2.24). It has the advantage over the direct evaluation of Eq. (6.33) that no curl operator has to be applied to $\underline{\underline{F}}_{\mathrm{p}}(t+\Delta t)$ to determine the new dislocation state, only the current slip gradients and the derivatives $\underline{\underline{\hat{e}}}_{,i}$ of the tensor exponentials are required. For the special case of plane deformation with two slip systems formula (6.36) can be further simplified. For this, the representation (5.24) for $\underline{\underline{\hat{\alpha}}}$ is used:

$$\begin{aligned} \underline{\underline{\hat{e}}} \cdot \underline{\underline{\hat{\alpha}}} &= (\hat{e}_{xx} \, \hat{\alpha}_{xz} + \hat{e}_{xy} \, \hat{\alpha}_{yz}) \, \underline{e}_x \otimes \underline{e}_z + (\hat{e}_{yx} \, \hat{\alpha}_{xz} + \hat{e}_{yy} \, \hat{\alpha}_{yz}) \, \underline{e}_y \otimes \underline{e}_z \, , \\ [\hat{e}][\hat{\alpha}] &= \begin{bmatrix} \hat{e}_{xx} & \hat{e}_{xy} & 0 \\ \hat{e}_{yx} & \hat{e}_{yy} & 0 \\ 0 & 0 & 1 \end{bmatrix} \begin{bmatrix} 0 & 0 & \hat{\alpha}_{xz} \\ 0 & 0 & \hat{\alpha}_{yz} \\ 0 & 0 & 0 \end{bmatrix} = \begin{bmatrix} 0 & 0 & \hat{e}_{xx} \, \hat{\alpha}_{xz} + \hat{e}_{xy} \, \hat{\alpha}_{yz} \\ 0 & 0 & \hat{e}_{yx} \, \hat{\alpha}_{xz} + \hat{e}_{yy} \, \hat{\alpha}_{yz} \\ 0 & 0 & 0 \end{bmatrix} . \end{aligned}$$

Using the abbreviation $\underline{\hat{V}}_i = \underline{\underline{\hat{e}}}^{\mathrm{T}} \cdot \underline{\widehat{\nabla}} \nu_i$ and $\underline{\underline{\hat{e}}}_{,i} \times \underline{\hat{V}}_i = \underline{\underline{\hat{e}}}_{,i} \cdot (-\underline{\underline{\epsilon}} \cdot \underline{\hat{V}}_i) =: \underline{\underline{\hat{e}}}_{,i} \cdot \underline{\underline{\hat{c}}}_i$ gives

$$\begin{aligned} \underline{\underline{\hat{e}}}_{,i} \times \underline{\hat{V}}_i = \underline{\underline{\hat{e}}}_{,i} \cdot \underline{\underline{\hat{c}}}_i &= (\hat{e}_{xx}^{,i} \, \hat{V}_{iy} - \hat{e}_{xy}^{,i} \, \hat{V}_{ix}) \, \underline{e}_x \otimes \underline{e}_z + (\hat{e}_{yx}^{,i} \, \hat{V}_{iy} - \hat{e}_{yy}^{,i} \, \hat{V}_{ix}) \, \underline{e}_y \otimes \underline{e}_z \, , \\ [\hat{e}_{,i}][\hat{c}_i] &= \begin{bmatrix} \hat{e}_{xx}^{,i} & \hat{e}_{xy}^{,i} & 0 \\ \hat{e}_{yx}^{,i} & \hat{e}_{yy}^{,i} & 0 \\ 0 & 0 & 0 \end{bmatrix} \begin{bmatrix} 0 & 0 & \hat{V}_{iy} \\ 0 & 0 & -\hat{V}_{ix} \\ -\hat{V}_{iy} & \hat{V}_{ix} & 0 \end{bmatrix} = \begin{bmatrix} 0 & 0 & \hat{e}_{xx}^{,i} \, \hat{V}_{iy} - \hat{e}_{xy}^{,i} \, \hat{V}_{ix} \\ 0 & 0 & \hat{e}_{yx}^{,i} \, \hat{V}_{iy} - \hat{e}_{yy}^{,i} \, \hat{V}_{ix} \\ 0 & 0 & 0 \end{bmatrix} . \end{aligned}$$

Thus the expression in braces in Formula (6.36) has only one xz and one yz component. The following multiplication from the right with $\underline{\underline{\hat{e}}}^{\mathrm{T}}$ therefore does not cause any change, so that finally a compact matrix form remains:

$$\begin{bmatrix} \hat{\alpha}_{xz}(t{+}\Delta t) \\ \hat{\alpha}_{yz}(t{+}\Delta t) \end{bmatrix} = \begin{bmatrix} \hat{e}_{xx} & \hat{e}_{xy} \\ \hat{e}_{yx} & \hat{e}_{yy} \end{bmatrix} \begin{bmatrix} \hat{\alpha}_{xz}(t) \\ \hat{\alpha}_{yz}(t) \end{bmatrix} - \sum_{i=1}^{2} \begin{bmatrix} \hat{e}_{xx}^{,i} & \hat{e}_{xy}^{,i} \\ \hat{e}_{yx}^{,i} & \hat{e}_{yy}^{,i} \end{bmatrix} \begin{bmatrix} \hat{V}_{iy}(t{+}\Delta t) \\ -\hat{V}_{ix}(t{+}\Delta t) \end{bmatrix},$$

$$\text{with} \quad \begin{bmatrix} \hat{V}_{ix} \\ \hat{V}_{iy} \end{bmatrix} = \begin{bmatrix} \hat{e}_{xx} & \hat{e}_{xy} \\ \hat{e}_{yx} & \hat{e}_{yy} \end{bmatrix}^{\mathrm{T}} \begin{bmatrix} F_{xx}^{\mathrm{e}} & F_{xy}^{\mathrm{e}} \\ F_{yx}^{\mathrm{e}} & F_{yy}^{\mathrm{e}} \end{bmatrix}^{\mathrm{T}} \begin{bmatrix} \nu_{i,x} \\ \nu_{i,y} \end{bmatrix}.$$

With the help of the analytical solution of the tensor exponential, the derivatives $[\hat{e}_{,i}]$ can be determined exactly in this special case (cf. Appendix A.2.1). This is especially important for the quality of the numerical derivation of the residual to determine the element stiffness matrix (6.21), where a double numerical differentiation is avoided. Furthermore, to determine $\underline{\underline{F}}_{\mathrm{p}}(t{+}\Delta t)$ as well as $\underline{\underline{\hat{\alpha}}}(t{+}\Delta t)$ the respective values are used at the previous time t. At the beginning $\underline{\underline{F}}_{\mathrm{p}}(t_0) = {}^0\underline{\underline{F}}_{\mathrm{p}}$ must be given as an initial condition, from which the initial state for $\underline{\underline{F}}(t_0) = \underline{\underline{I}}$ results:

$$\underline{\underline{F}}_{\mathrm{e}}(t_0) = {}^0\underline{\underline{F}}_{\mathrm{p}}^{-1} \;\rightarrow\; \underline{\underline{\hat{E}}}_{\mathrm{e}}(t_0) = \tfrac{1}{2}\left({}^0\underline{\underline{F}}_{\mathrm{p}}^{-\mathrm{T}} \cdot {}^0\underline{\underline{F}}_{\mathrm{p}}^{-1} - \underline{\underline{I}}\right), \; \underline{\underline{\hat{\alpha}}}(t_0) = \left(-{}^0\underline{\underline{F}}_{\mathrm{p}} \times \underline{\widetilde{\nabla}}\right) \cdot {}^0\underline{\underline{F}}_{\mathrm{p}}^{\mathrm{T}}. \tag{6.37}$$

Accordingly, $\underline{\underline{F}}_{\mathrm{p}}$ must be treated as a history variable, with ${}^0\underline{\underline{F}}_{\mathrm{p}}$ capturing the prehistory. From a physical point of view, this perspective is completely in line with the classical phenomenological theory of plasticity, according to which plastic (pre)history plays a role in material behavior [6, p. 436 ff.].

6.4 Global FE Algorithm

Before the actual solution algorithm begins, the domain is divided into suitable finite elements in a preprocessor step (meshing). In the present study, these are exclusively plane four-node quadrilateral elements. As shape functions $G_L(\chi, \eta)$, the well-known bilinear polynomials are used (cf. e. g. [3, p. 114]).

Now, given are the initial values of the history variables and, if applicable, the boundary values of the primary fields as well as the orientation of the crystal lattice. The basic domain Ω is divided into finite elements Ω_e. The process with the total duration T is evenly divided into time intervals Δt. With this everything on element level is ready to sketch the global FE algorithm for all Ω_e:

0. Specify the initial values $\underline{\underline{F}}_{\mathrm{p}}(\underline{r}, t = t_0)$ at quadrature points of all elements, thereby set the initial state (6.37).
1. Specify the nodal values $\underline{u}_L|_{\mathcal{B}}(t)$ and $\boldsymbol{\nu}_L|_{\mathcal{B}}(t)$ at the boundary $\mathcal{B}$.

2. Preset/calculate the deformation $\underline{\underline{F}}(t)$ at the current time/load step.
3. Integrate $\underline{\underline{F}}_{\text{p}}$ at all quadrature points of all elements according to Eq. (6.30), calculate $\underline{\underline{F}}_{\text{e}}$ and $\hat{\underline{\underline{E}}}_{\text{e}}$ (2.14) plus $\hat{\underline{\underline{\alpha}}}$ (6.36) at time $(t{+}\Delta t)$.
4. Evaluate the constitutive relations (cf. Sect. 5.3) in the lattice space to determine local kinetic quantities at the quadrature points of the elements.
5. Integrate the local kinetic quantities over the current element region according to Eq. (6.17) by means of Gaussian quadrature.
6. Calculate the element residual $[\overset{e}{r}]$ according to Eq. (6.18).
7. Determine the element stiffness matrix $[\overset{e}{K}_{LM}]$ (numerical differentiation).
8. Assemble the global stiffness matrix $[K]$ and the global FE equation system $[K][\Delta\varpi] = [r]$ taking into account the BCs from 1.
9. Solve the global FE equation system with the (discrete) Newton-Raphson method, thereby repeat the steps 2.–7. until $[r]$ is small enough.
10. Update the primary fields $\underline{u}(\underline{r}, t{+}\Delta t)$ and $\nu_i(\underline{r}, t{+}\Delta t)$.
11. Save the new values of the history variable $\underline{\underline{F}}_{\text{p}}$ and the state variable $\hat{\underline{\underline{\alpha}}}$.
12. Proceed to the next time/load step $t{+}\Delta t$: if $t < T$ go to 1., else: EXIT.

Mathematically speaking, the presented FE solution algorithm is an implicit method of 0th order. It shows similarities to the so-called γ-algorithm [7]. There, the plastic slip is used as DOF. The algorithm presented here can be regarded as a generalization of this, since it is designed for a geometrically nonlinear theory with slip *rates* ν_i as DOFs. The FE algorithm was implemented in a research code [8]. This FE code written in the programming language Java is characterized by a consistently object-oriented programming method and a modular structure, making it—as in the present case—flexibly extensible [8, 9]. Similar object-oriented FEM approaches and their advantages are presented elsewhere [10].

As it turned out, the modeling of the flow behavior has important implications on the FE algorithm: For the slip systems, the flow rule (3.37) or the corresponding dissipation potential (3.44) was chosen, so that there is no *hard* case distinction between elastic and elastic-plastic behavior (cf. Sect. 3.6). Applying a hard case distinction leads to an interface between plastically flowing ($\nu_i \neq 0$) and elastically deformed domains ($\boldsymbol{\nu} = 0$) within the material or a finite element (cf. also [5]). This discontinuity is accompanied by a jump in the material properties. For the correct description of the material behavior under hard flow criteria, the position and geometry of the interface would have to be modeled and tracked in its evolution. The use of plastic slip rates ν_i as degrees of freedom is hardly compatible with this, as jumps over the element cannot be represented with usual, continuous shape functions. Furthermore, the Gauss-Legendre quadrature scheme is only applicable for sufficiently smooth functions.[2] To avoid these problems another way was pursued here, following the basic idea of the phase field method (cf. e. g. [11]): There are

[2]There is a way out for this aspect via a domain-wise integration [9].

continuous transitions between plastically flowing and purely elastically deformed material domains. The blurring of sharp interfaces avoids discontinuities and the associated theoretical and numerical problems.

6.5 On the Thermodynamical Consistency and Steadiness of the FE Solution

A material model is *thermodynamically consistent* if it satisfies the Clausius-Planck inequality for all admissible motions and material parameter values. The evaluation of this inequality (cf. Sect. 3.5) results in the local condition (3.28), which must be fulfilled at every material point. Under certain assumptions, this led to a *strong* formulation of the multi-field problem of geometrically nonlinear CDT, which is thermodynamically consistent a priori. In addition, a *weak*, integral formulation (5.8) of the multi-field problem was obtained. The attributes *strong* and *weak* refer here to whether spatial derivatives of the Cauchy stresses $\underline{\underline{\sigma}}$ and Peach-Koehler forces $\underline{q}_1, \underline{q}_2$ are required in the formulation or not. Depending on the chosen formulation and (numerical) solution method, there are consequences for the fulfillment of the Clausius-Planck inequality.

One possibility is to choose the strong, local formulation of the multi-field problem as a starting point and to use the finite difference method (FDM) to solve it. This way was used in a study on the geometrically linear theory [12]. The advantage of this approach is that the thermodynamical consistency is maintained, despite the approximate nature of the solution. However, the disadvantage is that the FDM cannot easily be used for large deformations. The finite element method is better suited for this purpose, whereby the weak, integral formulation of the multi-field problem is chosen as the starting point. This approach offers the advantage that lower differentiability requirements can be placed on the functional basis, since no derivatives of the Cauchy stresses and Peach-Koehler forces are used. For the calculation of $\underline{\underline{\sigma}}$ (3.29) and $\underline{q}_i$ (3.30) only the first derivatives of the primary fields $\underline{u}$ and ν_1, ν_2 are needed. As the equilibrium conditions are integrated element by element, they are only fulfilled *in the weighted mean*[3] within the elements. Independent of the weak, integral formulation of the problem as basis of the FEM approximation solution, the strong, local formulation occurs during the evaluation of the thermodynamical consistency. This is problematic in relation to the continuity requirements of the primary fields. For the local verification of the fulfillment of the Clausius-Planck inequality stronger differentiability requirements are necessary than for the solution of the FEM equation system. Accordingly, a bilinear basis is not yet sufficient. A sufficiently high polynomial degree of the shape functions must be available, so that the necessary derivatives of the Cauchy stresses and Peach-Koehler forces within the elements can be calculated. Another possibility is to introduce these derived quantities—here $\text{div}(\underline{\underline{\sigma}})$, $\text{div}(\underline{q}_1)$, $\text{div}(\underline{q}_2)$—as additional DOFs with their own shape functions.

[3] In the Bubnov-Galerkin method the weighting factors correspond to the shape functions.

However, this might not be enough: Sufficient continuity should also be guaranteed across element boundaries. In the present work $\underline{u}(\underline{r})$ and $\nu_i(\underline{r})$ are continuous, but not continuously differentiable at the element edges due to the FE approach (6.2). While $\underline{\underline{F}}_\mathrm{p}(\boldsymbol{\nu})$ according to the Eq. (2.12) is thus continuous at the element boundaries, the deformation $\underline{\underline{F}}(\mathrm{grad}(\underline{u}))$ according to Eq. (6.29) can show jumps there. This also makes $\underline{\underline{F}}_\mathrm{e} = \underline{\underline{F}} \cdot \underline{\underline{F}}_\mathrm{p}^{-1}$ discontinuous at the element boundaries. Accordingly, the equality (2.18) is no longer fulfilled at the element edges.[4] Since $\hat{\underline{\underline{\alpha}}}$ also depends on $\mathrm{grad}(\nu_i)$, the dislocation state—just like the deformation—can have jumps at the element boundaries. To avoid this, specific, more complicated finite elements exist [13], which allow a globally continuously differentiable solution. However, as the presented theory only invokes $\mathrm{Curl}(\underline{\underline{F}}_\mathrm{p})$, it is not necessary that the solution for the plastic distortion field is continuous such that $\mathrm{Grad}(\underline{\underline{F}}_\mathrm{p})$ exists everywhere in the domain Ω. Therefore, the requirements on the mathematical solution space can be weakened, resulting in a so-called $H_1(\mathrm{Curl}; \Omega)$-space [14, 15]. With $\underline{\underline{F}}_\mathrm{p}$ beeing a function in this space, only its "tangential component" $\mathrm{Curl}(\underline{\underline{F}}_\mathrm{p})$ needs to be continuous across the element boundaries. Then, according to Eq. (2.18), the dislocation density state $\hat{\underline{\underline{\alpha}}}$ is continuous as well. As in FE methods for Maxwell's theory of electromagnetism [14], Nédélec's finite elements can be used to fulfill these continuity requirements [15–17].

In summary, it is stated that the FE approximation solution can locally violate the Clausius-Planck inequality. Just as the field equations are only fulfilled on average, the same is to be expected for the thermodynamical consistency. However, this local disregard of the Clausius-Planck inequality must not have any global consequences. To ensure this, the power dissipated per element can be evaluated. According to the assumption (3.32) this depends here only on the slip rates ν_1, ν_2 and dissipative resolved shear stresses κ_1, κ_2, so that the following can be demanded:

$$\overset{e}{\mathcal{D}} = \int\limits_{V_e} [\nu]^\mathrm{T}[\kappa]\,\mathrm{d}V = \int\limits_{V_e} [\overset{e}{\dot{w}}]^\mathrm{T}[G_\nu][\kappa]\,\mathrm{d}V \overset{!}{\geq} 0 \quad \text{with} \quad [\kappa] = \begin{bmatrix} \kappa_1 \\ \kappa_2 \end{bmatrix} = \Lambda \begin{bmatrix} \partial d/\partial\nu_1 \\ \partial d/\partial\nu_2 \end{bmatrix}.$$

Accordingly, the fulfillment of the flow rules derived from the dissipation potential d determines the thermodynamical consistency (cf. Sect. 3.7). In this way it can be guaranteed for the following simulation results that the energy dissipated in the element and the entropy produced with it will not become negative. Since the problem is basically a discretization error, it can be reduced by finer meshing. In addition, the use of energy-entropy-consistent FE integrators [18] could help.

[4] This problem exists even if the dislocation state $\hat{\underline{\underline{\alpha}}}$ is determined by Eq. (6.36).

References

1. Akpama, H.K., Bettaieb, M.B., Abed-Meraim, F.: Numerical integration of rate-independent BCC single crystal plasticity models: comparative study of two classes of numerical algorithms. Int. J. Numer. Methods Eng. **108**(5), 363–422 (2016)
2. Szabó, B.A., Babuska, I.: Finite Element Analysis. Wiley, New York (2008)
3. Wriggers, P.: Nonlinear Finite Element Methods. Springer, Berlin (2008)
4. Gurtin, M.E.: A finite-deformation, gradient theory of single-crystal plasticity with free energy dependent on the accumulation of geometrically necessary dislocations. Int. J. Plast. **26**(8), 1073–1096 (2010)
5. Donner, H.: FEM-basierte Modellierung stark anisotroper Hybridcord-Elastomer-Verbunde. Ph.D. thesis, Technische Universität Chemnitz (2017)
6. Haupt, P.: Continuum Mechanics and Theory of Materials, vol. 2. Springer, Berlin (2002)
7. Klusemann, B., Svendsen, B., Bargmann, S.: Analysis and comparison of two finite element algorithms for dislocation density based crystal plasticity. GAMM-Mitteilungen **36**(2), 219–238 (2013)
8. Baitsch, M., Hartmann, D.: Piecewise polynomial shape functions for hp-finite element methods. Computer Methods Appl. Mech. Eng. **198**(13–14), 1126–1137 (2009)
9. Baitsch, M., Le, K.C., Tran, T.M.: Dislocation structure during microindentation. Int. J. Eng. Sci. **94**, 195–211 (2015)
10. Gori, L., Penna, S.S., Pitangueira, R.L.: A computational framework for constitutive modelling. Comput. Struct. **187**, 1–23 (2017)
11. Bulatov, V., Cai, W.: Computer Simulations of Dislocations (Oxford Series on Materials Modelling). Oxford University Press, USA (2006)
12. Silbermann, C.B., Ihlemann, J.: Geometrically linear continuum theory of dislocations revisited from a thermodynamical perspective. Arch. Appl. Mech. **88**(1–2), 141–173 (2017)
13. Fischer, P., Klassen, M., Mergheim, J., Steinmann, P., Müller, R.: Isogeometric analysis of 2D gradient elasticity. Comput. Mech. **47**(3), 325–334 (2010)
14. Larson, M.G., Bengzon, F.: Electromagnetics, pp. 327–354. Springer, Berlin (2013)
15. Wieners, C., Wohlmuth, B.: A primal-dual finite element approximation for a nonlocal model in plasticity. SIAM J. Numer. Anal. **49**(2), 692–710 (2011)
16. Panteghini, A., Bardella, L.: On the role of higher-order conditions in distortion gradient plasticity. J. Mech. Phys. Solids **118**, 293–321 (2018)
17. Adedoyin, A.A., Enakoutsa, K., Bammann, D.J.: An assessment of the evolving microstructural model of inelasticity coupled with dislocation- and disclination-based incompatibilities. J. Eng. Mat. Technol. **141**(4) (2019)
18. Krüger, M.: Energie-Entropie-konsistente Zeitintegratoren für die nichtlineare Thermoviskoelastodynamik. Ph.D. thesis, Universität Siegen (2012)

Chapter 7
FE Simulation Results

Abstract This chapter presents numerical solutions of the initial boundary value problem for a continuously dislocated single crystal under plane shear deformation. To this end, the in-house finite element simulation code explained in the preview chapter is adopted. The simulation results are discussed in the light of the theory of complex, pattern forming systems. Additionally, the convergence behavior of the FE solution is examined and numerical challenges are highlighted.

7.1 Simulation Setting

In the following simulations, a rectangular domain of the area $A = L_x \cdot L_y$ with the boundary $\mathcal{B}$ is considered, which is discretized with a number of N_e quadratic finite elements (cf. Fig. 7.1, left). The angle φ describes the rotation between the global sample coordinate system and the crystal coordinate system. There are two initially orthogonal slip systems (SS 1 and 2). For the time integration, N_t time intervals of the same size Δt are used. Furthermore, the initial value $\underline{\underline{F}}_{\mathrm{p}}(0) = \underline{\underline{I}}$ is assumed always (no plastic prehistory) and a bilinear functional basis is selected. To verify the implementation the special cases from Sect. 4 are considered first.

7.2 Plausibility Check Based on Single Crystal Elasticity

The simplest case of single crystal elasticity (4.1) is suitable for studying the effects of cubic anisotropy on large deformations. As plausibility check, the investigation of the material symmetry is particularly suitable. To generate different reference configurations, the orientation of the crystal coordinate system is varied according to the Relation (5.15). This shows that $\varphi = \varphi_0$ and $\varphi = \varphi_0 + \pi/2$ lead to an iden-

---

Electronic supplementary material The online version of this chapter (https://doi.org/10.1007/978-3-030-63696-8_7) contains supplementary material, which is available to authorized users.

C. B. Silbermann et al., *Introduction to Geometrically Nonlinear Continuum Dislocation Theory*, SpringerBriefs in Continuum Mechanics,
https://doi.org/10.1007/978-3-030-63696-8_7

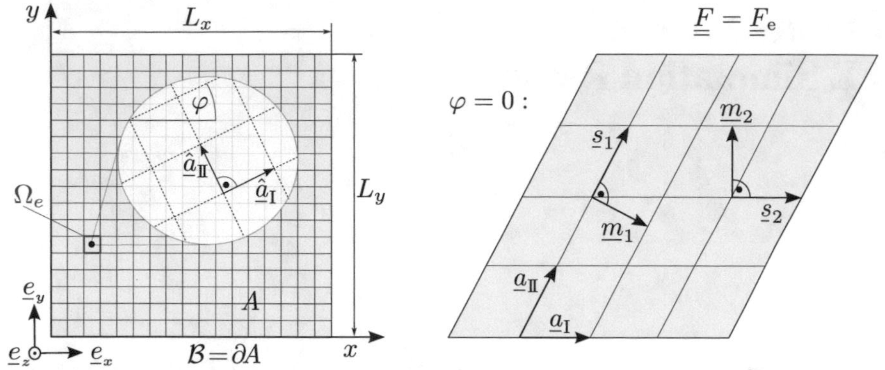

Fig. 7.1 *Left* Basic domain of the area A with quadratic finite elements Ω_e and underlying crystal lattice with orientation φ; *Right* Elastic shearing with initial orientation $\varphi = 0$: shown are current lattice vectors and slip systems

tical material behavior. Thus, rotations of the crystal by 90° have no effect, which corresponds to the expected symmetry.

Next, the compliance with the transformation rules (2.5) is checked using the example of homogeneous elastic shear. Due to the lattice deformation, the lattice vectors in the current configuration are no longer perpendicular to each other and the slip systems are no longer orthogonal (cf. Fig. 7.1, right). The slip planes are also rotated, with the slip directions being in the corresponding plane. Accordingly, the effect of the elastic deformation on the slip kinematics is correctly reproduced.

7.3 Homogeneous Shearing of Elastoviscoplastic Single Crystals

The next special case is the standard model of single crystal elastoviscoplasticity (4.2). A plausibility analysis of the slip system activity is carried out with a simple shear test. Due to the prerequisite of a cubic primitive crystal lattice it is possible to model the shear test as a plane problem with *two* active slip systems. As with single crystal elasticity, elastic anisotropy is taken into account.

The shear test is conducted quasistatically ($\ddot{\underline{u}} \to \underline{0}$) and displacement-controlled by setting Dirichlet-BCs for $\underline{u}$ at the entire boundary $\mathcal{B}$. The total deformation is

$$\underline{\underline{F}} = \underline{\underline{I}} + \gamma(t)\, \underline{e}_x \otimes \underline{e}_y\,. \tag{7.1}$$

For the slip rates ν_1, ν_2 the natural BCs $\pi_1 = \pi_2 = 0$ according to representation (5.14c) were applied everywhere. Accordingly, dislocations can leave the crystal unhindered (and thus disappear), allowing homogeneous deformation. Furthermore, the slip system orientation $\varphi = 0$ is chosen, so that crystal and sample coordinate system coincide (cf. Fig. 7.1, right). According to approach (5.16) it then holds

$$\hat{\underline{\underline{L}}}_{\text{p}} = \nu_1(x, y, t)\, \underline{e}_y \otimes \underline{e}_x + \nu_2(x, y, t)\, \underline{e}_x \otimes \underline{e}_y \,. \tag{7.2}$$

Obviously, the activity of SS 2 is sufficient to carry the total deformation, whereas SS 1 may remain inactive. The special case of single-slip from Sect. 4.3 provides an analytical preview. According to the Formula (4.3), if SS 1 is the only slip system and given the IC $\underline{\underline{F}}_{\text{p}}(0) = \underline{\underline{I}}$ it follows

$$\underline{\underline{F}}_{\text{p}}(t) = \underline{\underline{I}} + \beta_1\, \underline{e}_y \otimes \underline{e}_x \quad \text{with} \quad \beta_i = \int_{t=0}^{t=t} \nu_i(t)\,\text{d}t \,, \tag{7.3}$$

whereby the elastic part of the deformation according to $\underline{\underline{F}}_{\text{e}} = \underline{\underline{F}} \cdot \underline{\underline{F}}_{\text{p}}^{-1}$ reads

$$\underline{\underline{F}}_{\text{e}}(t) = \underline{\underline{I}} + \gamma\, \underline{e}_x \otimes \underline{e}_y - \beta_1\, \underline{e}_y \otimes \underline{e}_x - \gamma\beta_1\, \underline{e}_x \otimes \underline{e}_x \,. \tag{7.4}$$

If SS 2 is the only active SS then Eq. (7.3) needs the substitution $\beta_1(t)\, \underline{e}_y \otimes \underline{e}_x \rightarrow \beta_2(t)\, \underline{e}_x \otimes \underline{e}_y$. In a completely analogous way, the elastic distortion then follows

$$\underline{\underline{F}}_{\text{e}}(t) = \underline{\underline{I}} + (\gamma - \beta_2)\, \underline{e}_x \otimes \underline{e}_y \,. \tag{7.5}$$

The different behavior due to the activity of SS 1 or 2 becomes particularly clear in the corresponding elastic distortions according to the Formula (2.14):

$$\left.\begin{bmatrix} \hat{E}^{\text{e}}_{xx} \\ \hat{E}^{\text{e}}_{yy} \\ 2\hat{E}^{\text{e}}_{xy} \end{bmatrix}\right|_{\beta_2=0} = \begin{bmatrix} -\beta_1(\gamma - (1+\gamma^2)\beta_1/2) \\ \frac{1}{2}\gamma^2 \\ \gamma - (1+\gamma^2)\beta_1 \end{bmatrix} , \quad \left.\begin{bmatrix} \hat{E}^{\text{e}}_{xx} \\ \hat{E}^{\text{e}}_{yy} \\ 2\hat{E}^{\text{e}}_{xy} \end{bmatrix}\right|_{\beta_1=0} = \begin{bmatrix} 0 \\ \frac{1}{2}(\gamma - \beta_2)^2 \\ (\gamma - \beta_2) \end{bmatrix} . \tag{7.6}$$

This results in the corresponding stress state for $\varphi = 0$ according to the stress strain relationship (3.9) or the matrix form (5.22):

$$\begin{bmatrix} \widehat{T}^{\text{e}}_{xx} \\ \widehat{T}^{\text{e}}_{yy} \\ \widehat{T}^{\text{e}}_{xy} \end{bmatrix} = \begin{bmatrix} \varkappa\, \hat{E}^{\text{e}}_{xx} + \lambda\, \hat{E}^{\text{e}}_{yy} \\ \lambda\, \hat{E}^{\text{e}}_{xx} + \varkappa\, \hat{E}^{\text{e}}_{yy} \\ \mu\, \hat{E}^{\text{e}}_{xy} + \mu\, \hat{E}^{\text{e}}_{xy} \end{bmatrix} . \tag{7.7}$$

The elastic shear strains and stresses can be kept small by the sole activity of SS 1 as well as SS 2—paradoxically with a lower slip $\beta_1 = \frac{\gamma}{1+\gamma^2}$ in SS 1 compared to $\beta_2 = \gamma$ in SS 2. However, the rotation by SS 1 contrary to the given deformation (7.1) causes considerable elastic normal strains and stresses in x- and y-direction, which is not the case with SS 2 at all (or much less). The shear stresses resolved in the slip systems according to Formula (3.34) with $\hat{C}^{\text{e}}_{ab} = 2\hat{E}^{\text{e}}_{ab} + \delta_{ab}$ are

$$\tau_1 = \underline{s}_1 \cdot \underline{\underline{\sigma}} \cdot \underline{m}_1 = \underline{e}_y \cdot \tfrac{1}{j}\, \hat{\underline{\underline{C}}}_{\text{e}} \cdot \widehat{\underline{\underline{T}}}_{\text{e}} \cdot \underline{e}_x = \hat{C}^{\text{e}}_{yx}\widehat{T}^{\text{e}}_{xx} + \hat{C}^{\text{e}}_{yy}\widehat{T}^{\text{e}}_{yx} \,, \tag{7.8}$$

$$\tau_2 = \underline{s}_2 \cdot \underline{\underline{\sigma}} \cdot \underline{m}_2 = \underline{e}_x \cdot \tfrac{1}{j}\, \hat{\underline{\underline{C}}}_{\text{e}} \cdot \widehat{\underline{\underline{T}}}_{\text{e}} \cdot \underline{e}_y = \hat{C}^{\text{e}}_{xx}\widehat{T}^{\text{e}}_{xy} + \hat{C}^{\text{e}}_{xy}\widehat{T}^{\text{e}}_{yy} \,. \tag{7.9}$$

The analytical consideration for homogeneous single-slip can be completed by determining the evolution $\beta_i(t)$. With power law (3.37) as flow rule it follows

$$\beta_i(t) = \int_{t=0}^{t=t} \nu_i(t)\,\mathrm{d}t = \int_{t=0}^{t=t} \frac{\tau_{\mathrm{cr}}}{\eta_{\mathrm{p}}} \left(\frac{\tau_i(t)}{\tau_{\mathrm{cr}}}\right)^p \mathrm{d}t\ . \tag{7.10}$$

Since $\tau_i(t)$ is a function $\tau_i(\beta_i(t), \gamma(t))$, (7.10) is an integral equation. Taking the time derivative the integration is omitted and a nonlinear ordinary first order differential equation for the determination of β_i is obtained. A closed analytical solution is only possible in special cases: If a monotonous course of $\gamma(t)$ is assumed as well as rate-independent slip ($p \to \infty$) in one active SS 2, the analytical solution is composed as follows:

$$\beta_2(t) = \begin{cases} 0 & : \tau_2(t) < \tau_{\mathrm{cr}} \\ \gamma(t) - \gamma_{\mathrm{cr}} & : \tau_2(t) = \tau_{\mathrm{cr}} \end{cases} \quad \text{with } \tau_2 = \mu(\gamma - \beta_2) + \frac{\kappa}{2}(\gamma - \beta_2)^3 \text{ and } \gamma_{\mathrm{cr}} \approx \frac{\tau_{\mathrm{cr}}}{\mu}\ .$$

The shear stress $\tau_2(t)$ increases steadily until γ_{cr} is reached. From then on SS 2 becomes active such that τ_2 always remains τ_{cr}. This results in a kink in the course of $\tau_2(t)$ as a transition from elastic to ideal plastic behavior.

Due to the nonlinear kinematics (cf. Sect. 2.4) the multi-slip behavior with SS 1 and 2 *cannot* be represented analytically by a superposition of two single-slip processes. Accordingly, numerical simulations are performed using the parameters from Table 7.1, which are plausible, but here not assigned to a certain metal.

The anisotropy parameter (3.8) is thus $A = 2$, which impedes shear deformations (shear stiffening) compared to the isotropic case with $A = 1$.

Due to the initial orthogonality, both slip systems initially have the same shear stress, whereby both become active. Due to the elastic lattice deformation SS 1 rotates, but SS 2 does not (cf. Fig. 7.1, right). This leads to a minimal preference of SS 2. As a result of the activity of SS 2, this preference is increased and the plastic deformation takes place (almost) exclusively in SS 2. For this to take effect, the exponent p in flow rule (7.10) must be sufficiently high (cf. Fig. 7.2). For $p \to \infty$ rate *in*dependence is reached, which can be solved here with the algorithm from Appendix A.2.2. After a short activity SS 1 remains permanently inactive. This is not the case for rate dependence: The reduced but persistent activity of SS 1 results in an increasing "hardening" (cf. Fig. 7.2). The rate dependence is illustrated by looking at two different process times. For $p \to 1$, the viscous, rate-dependent behavior becomes more and more noticeable, resulting in resolved shear stresses $|\tau|$ above or

Table 7.1 Material parameters for single crystal elastoviscoplasticity

$\tilde{\varrho}/\frac{g}{cm^3}$	$\varkappa$/GPa	λ/GPa	μ/GPa	η_{p}/MPa s	τ_{cr}/MPa	b/μm	k
1.0	200	100	100	$5 \cdot 10^{-3}$	10	$3 \cdot 10^{-4}$	0

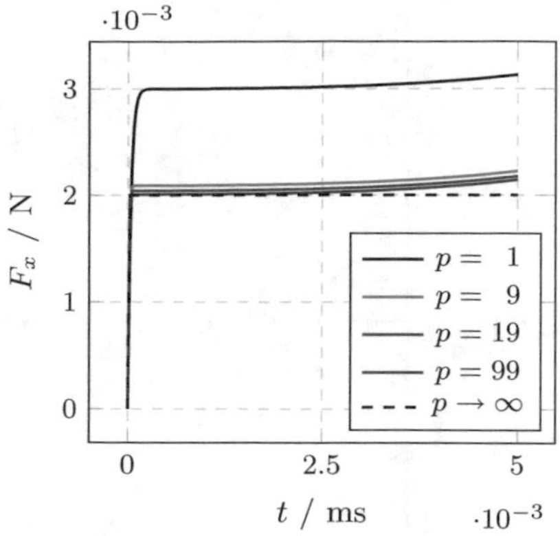

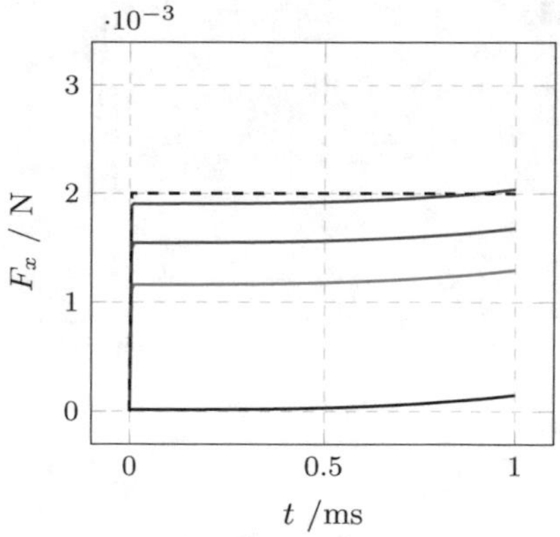

Fig. 7.2 Shear force F_x over time for shearing until $\gamma = 0.03$ at two different process times. Rate independence ($p \to \infty$) yields $F_{cr} = \tau_{cr} \cdot 200\ \mu m^2 = 2$mN

below τ_{cr} according to the selected process time and viscosity (cf. Fig. 7.2 left and right). For $p \to \infty$ always $|\tau| = \tau_{cr}$ applies.

Finally, the non-convexity of the free energy as described by [1] as well as its effect is investigated by means of the homogeneous shear test. Again, the exclusive activity of SS 2 is assumed, whereby the initial lattice orientation φ should now be variable according to the Formula (5.14). This leads to

$$\underline{\underline{F}}_p(t) = \underline{\underline{I}} + \beta_2\, \hat{\underline{s}}_2 \otimes \tilde{\underline{m}}_2 = \underline{\underline{I}} + \beta_2 \left(\cos\varphi\, \underline{e}_x + \sin\varphi\, \underline{e}_y\right) \otimes \left(-\sin\varphi\, \underline{e}_x + \cos\varphi\, \underline{e}_y\right) .$$

With shear deformation (7.1) and $\underline{\underline{F}}_e = \underline{\underline{F}} \cdot \underline{\underline{F}}_p^{-1}$ plus the notation $\beta_2 = \beta$ it follows

$$\begin{bmatrix} F^e_{xx} & F^e_{xy} \\ F^e_{yx} & F^e_{yy} \end{bmatrix} = \begin{bmatrix} 1 + \beta\left(\frac{1}{2}\sin(2\varphi) + \gamma\sin^2\varphi\right) & \gamma - \beta\left(\frac{\gamma}{2}\sin(2\varphi) + \cos^2\varphi\right) \\ -\beta\sin^2\varphi & 1 - \frac{\beta}{2}\sin(2\varphi) \end{bmatrix} \tag{7.11}$$

If β is set opposite to the control parameter γ such that $[F_e] \overset{!}{=} \begin{bmatrix} -\cos(2\varphi) & -\sin(2\varphi) \\ \sin(2\varphi) & -\cos(2\varphi) \end{bmatrix}$ becomes a rotation matrix, there is no more elastic lattice strain, but only a lattice rotation $[R_e]$. This is fulfilled if the following applies [2]:

$$\gamma = -2\cot\varphi \quad \text{and} \quad \beta = 2\cot\varphi . \tag{7.12}$$

The significance of these relationships is explained using the initial orientation $\varphi = -\frac{\pi}{3} = -60°$ with $\cot\varphi = -3^{-\frac{1}{2}} \approx -0.577$: If the externally controlled shear γ reaches the value $-2\cot\varphi \approx 1.15$, a plastic slip $\beta \approx -1.15$ would lead to an undistorted, merely rotated lattice. Thus the strain energy $\hat{\phi}_e$ (3.4) has a second minimum at $\beta = -\gamma \approx -1.15$ in addition to the point $\beta = \gamma = 0$. Whether and when this state occurs depends on the plastic flow resistance of the material. When considering the

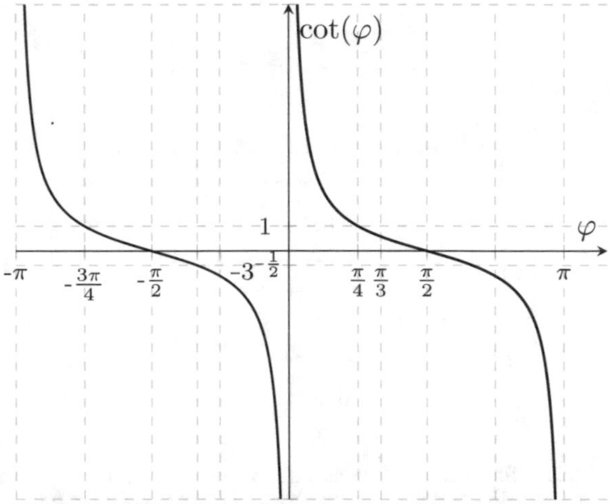

Fig. 7.3 Function $\cot(\varphi)$ in the range $(-\pi, \pi) \times (-10, 10)$

function $\cot(\varphi)$ (cf. Fig. 7.3) it becomes clear that the non-convexity can be lost depending on the initial orientation φ. This occurs e. g. at $\varphi = -\frac{\pi}{2}$, where both minima coincide at the origin; or at $\varphi = 0$, where the second minimum moves infinitely away from the origin. Between these extreme cases, e. g. in the interval $(-\frac{\pi}{2}, 0)$, the second minimum is reached if the given shear is sufficiently large.

If the BCs allow a homogeneous deformation and $\underline{\underline{F}}_{\mathrm{e}} = \underline{\underline{R}}_{\mathrm{e}} = \text{const.}$ applies, no dislocations are geometrically necessary, so that neither strain nor dislocation energy is stored in the material. With an *inhomogeneous* (shear) deformation under plastic single or double slip, the tendencies

$$\underline{\underline{F}}_{\mathrm{e}}(\gamma, \beta, \varphi) \rightarrow \underline{\underline{R}}_{\mathrm{e}}(\varphi) \quad \text{resp.} \quad \underline{\underline{F}} \cdot \underline{\underline{F}}_{\mathrm{p}}^{-1}(\nu_1, \nu_2, \varphi) \rightarrow \underline{\underline{R}}_{\mathrm{e}}(\varphi) \tag{7.13}$$

can occur locally, with the slip rates acting as order parameters. The slip processes are then energetically favorable, if $\underline{\underline{F}}_{\mathrm{e}}$ is striving against a pure lattice rotation $\underline{\underline{R}}_{\mathrm{e}}$—provided the (local) conditions allow it. Since an analytical evaluation of these conditions seems impossible, numerical investigations using FEM follow.

7.4 Heterogeneous Shearing and Simulation of Subgrain Formation

After the basic plausibility considerations on the basis of homogeneous deformations, an aspect of the so-called *grain refinement* shall now be investigated. As already described in detail [3] this is a dislocation-induced structure formation process that can be roughly divided into two stages [4]: The evolution of dislocation cells in stage (i) has already been investigated and modeled (c.f. e. g. [3, 5]). As plastic deformation proceeds, crystallographic orientation differences form between these cells and the cell walls transform into subgrain boundaries (stage ii). In the present paper it is assumed that these orientation differences result from an accumulation of GNDs. It is not yet clear whether the formation of orientation differences takes place continuously or abruptly. The assumption of a sudden formation could be explained in such a way that a complete cell rotation avoids an energetically unfavorable lattice distortion within the cell. In order to be able to simulate this presumed effect, the elastic distortion of the crystal lattice and its feedback to plastic slip must be taken into account. For this reason, the use of geometrically nonlinear CDT is necessary. In comparison to purely phenomenological plasticity [6] and also to geometrically linear CDT [7] a new degree of freedom exists here: the rotation of the crystal lattice. This opens up new deformation mechanisms, whereby a basic classification is possible on the basis of the homogeneity during the deformation of a crystallite:

(a) Due to a homogeneous strain and rotation, the lattice (more precisely, the lattice vectors $\underline{a}_K$) remains uniform in the entire crystallite. Plastic slip can take place unhindered, and no dislocations are *geometrically* necessary.
(b) For an inhomogeneous deformation of the crystallite, the lattice vectors $\underline{a}_K(\underline{r})$ become spatially dependent. Dislocations remain in the crystal where they are geometrically necessary.

Case (a) was already discussed in the previous section, where it was shown that a homogeneous lattice rotation can minimize the free energy of the crystal. Case (b) implies that the inhomogeneous deformation is composed of *regionally* homogeneous states. In between, GNDs accumulate. If the energy stored therein is sufficiently low, such a subgrain structure becomes energetically advantageous compared to an unstructured heterogeneous state of the crystallite.

To verify these considerations, a crystallite in the form of a polygon is modeled. The edges correspond to idealized grain boundaries, which represent impermeable interfaces for dislocations (BC $\nu_i\big|_{\mathcal{B}} = 0$). As a result, only elastic deformations are possible at $\mathcal{B}$. This excludes a homogeneous, undistorted lattice rotation with $\underline{\underline{F}}_{\mathrm{e}} = \underline{\underline{R}}_{\mathrm{e}}$ as described in Sect. 7.3 and heterogeneous deformation must occur.

Specifically, shearing of a quadratic crystallite of the size $L_x \equiv L_y = L = 200\ \mu\mathrm{m}$ is now considered and the domain is discretized with $N_e = 99 \cdot 99$ elements (cf. p. 57). The process time is $T = 10^{-5}$s, and $N_t = 150$ time intervals of the same size are used. The experiment is carried out quasistatically and displacement-controlled by setting Dirichlet-BCs for $\underline{u}$ at the entire boundary $\mathcal{B}$ and monotonous deformation up to the shear value $\gamma = 0.5$. The material parameters are first chosen according to Table 7.1, except $k = 10^{-4}$, and the initial orientation is $\varphi = -\frac{\pi}{3} = -60°$. Further, flow rule (3.36) is applied and only SS 2 is active.

For small to moderate externally prescribed shear values, the results are similar to those of linear theory: GNDs accumulate near $\mathcal{B}$, whereas the interior remains almost dislocation-free and deforms—largely homogeneous—in an elastic-plastic manner (cf. Fig. 7.4 top left). Above the shear value $\gamma \approx 0.2$ this behaviour changes. First, the

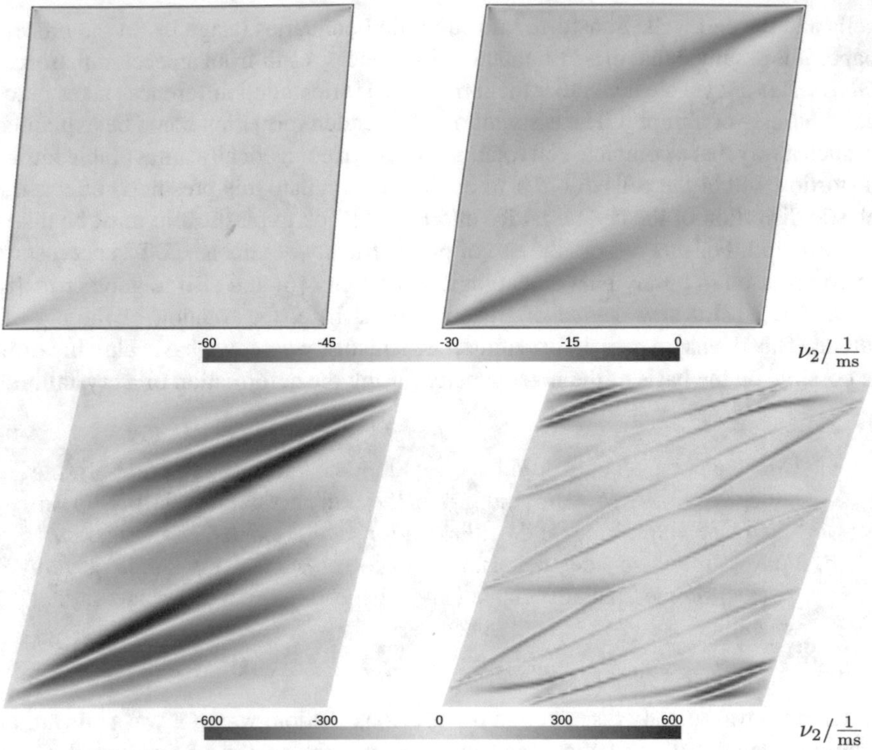

Fig. 7.4 Shear with only SS 2 being active, initial orientation $\varphi = -60°$: Depicted is the slip rate in SS 2 to four different phases of deformation

plastic slip is localized in bands (cf. Fig. 7.4 top right) similar to shear bands. The slip rates in the rest of the crystallite continue to decrease to zero. With further localization of the plastic deformation in the shear bands, the slip in the adjacent regions finally even runs in the opposite direction (cf. Fig. 7.4 bottom left). Due to these opposing slip processes, regions of different, but largely uniform lattice orientation develop. In between, GNDs accumulate and form a deformation boundary. Thus, the simulation results show the expected region-wise homogeneous solution (cf. Fig. 7.4 bottom right).

In addition to the evaluation of the current slip rates, the consideration of the resulting elastic-plastic deformation is particularly informative. From the rotational part $\underline{\underline{R}}_{\text{e}}$ of the elastic distortion a scalar rotation angle ϑ can be determined by means of $I_1(\underline{\underline{R}}_{\text{e}}) = 2\cos(\vartheta)$. The angle ϑ fully describes the plane lattice rotation around the z-axis. As can be seen from Fig. 7.5, after shearing up to $\gamma = 0.5$ homogeneously rotated lattice regions were formed (including six bands with negative lattice rotation). As Fig. 7.6 shows, deformation boundaries have formed in SS 2.[1] These consist

[1] The dislocation density in the inactive SS 1 is zero as expected and is therefore not depicted.

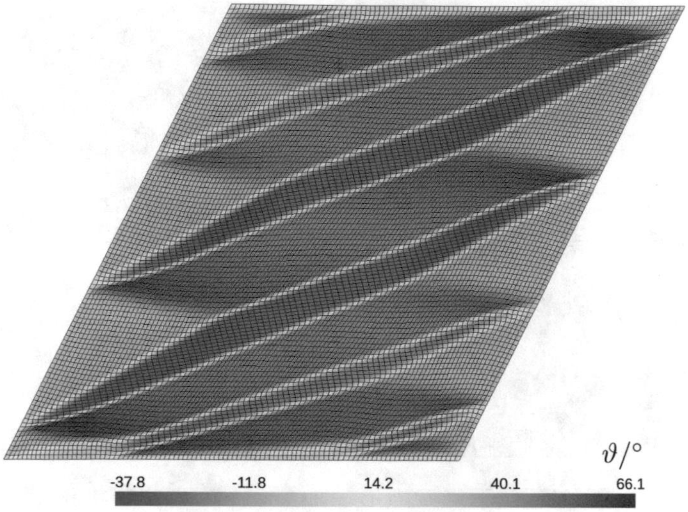

Fig. 7.5 Shear with only SS 2 being active, initial orientation $\varphi = -60°$: On the basis of the lattice rotation ϑ after shearing up to $\gamma = 0.5$ the differently rotated regions in the crystallite can be clearly identified as subgrains. A corresponding animation of the evolution can be found at Supplementary material 1

of concentrated GNDs of the same sign. Physically this corresponds to piled edge dislocations with the same tangent and Burgers vector. Dislocation walls constructed in this way form tilt boundaries [8, p. 40], which is often observed experimentally [9, p. 175].[2] However, these tilt boundaries are usually low-angle boundaries, which is related to the fact that the distance between real, discrete dislocations cannot become arbitrarily small. For a symmetrical tilt boundary, the mean vertical distance D of the piled edge dislocations results from $2D = b\sin(\Delta\vartheta/2)$ [9, p. 175]. For $\Delta\vartheta \geq 60°$ (cf. Fig. 7.5) $D \leq b$ and the individual dislocations can hardly be distinguished from each other. The high-angle boundaries formed in this way (cf. Fig. 7.6) therefore go beyond a mere piling of edge dislocations and represent a new two-dimensional lattice defect that can be captured by the theory without having been explicitly considered.

The simulation was repeated for different dislocation energies, with the weighting factor k varying between 0 and 10^2. As it turns out, the effect is small in the physically meaningful range for $k \lesssim 10^{-1}$. The dislocation energy has the effect of an interface energy, where the special form (3.13) limits admissible incompatibilities (i. e. the dislocation density). The fact that almost identical results occur for $k = 0$ as for $k = 10^{-4}$ proves that the instability is not caused by the consideration of the dislocation energy. Nevertheless, structure formation is dislocation-induced in the sense that it becomes possible only through lattice incompatibilities due to GNDs: they play

[2]The also frequently observed arrangement of dipoles (dislocations of different signs) at 45° cannot be reproduced with theories that only take GNDs into account [10].

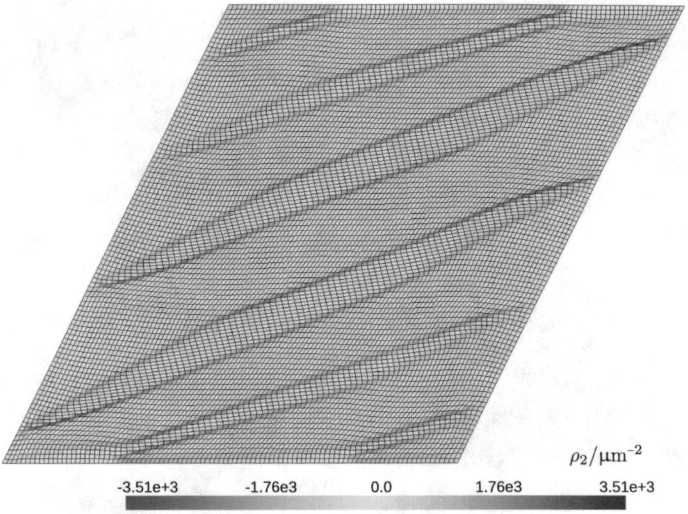

Fig. 7.6 Shear with only SS 2 being active, initial orientation $\varphi = -60°$: Depicted is the signed dislocation density in SS 2 after shearing up to $\gamma = 0.5$. Walls of GNDs form tilt boundaries and enclose dislocation-free subgrains

the decisive role in that they enable an inhomogeneous lattice rotation at all. Here k controls the sharpness or width of the deformation boundaries due to the repulsive Peach-Koehler forces between dislocations of the same sign (cf. Remark 3.1, p. 22). At (unphysically) high values $k \gtrsim 10^2$ the formation of the described deformation boundaries becomes energetically so "expensive" that instead of 12 only 4 deformation boundaries can be formed (cf. Figs. 7.6 with 7.7). These are clearly expanded and the maximum values of the dislocation density are lower.

Furthermore, the simulation was performed with $k = 10^{-4}$ for a domain 100 times smaller with the edge lengths $L_x = L_y = 20$ μm. It is remarkable that the subgrain structure, which was formed during shearing, remained topologically the same (cf. Fig. 7.8). Accordingly, it is independent of the crystallite size. Since a 100 times larger crystallite produces the same subgrain structure (cf. Fig. 7.6), the influences of the boundaries are obviously very long-range and represent an essential influencing factor of the structure formation. A closer look at the deformation boundaries further shows that they appear wider in the smaller specimen and have a maximum dislocation density that is 10 times higher. Two conclusions can be drawn from these results:

1. The subgrain structure results from the instability due to the non-convexity of the elastic energy and has no internal length scale (scaling invariance).
2. The width of the deformation boundaries is influenced by the internal length scale $\ell = (b\rho_\mathrm{s})^{-1}$, which arises from the dislocation energy.

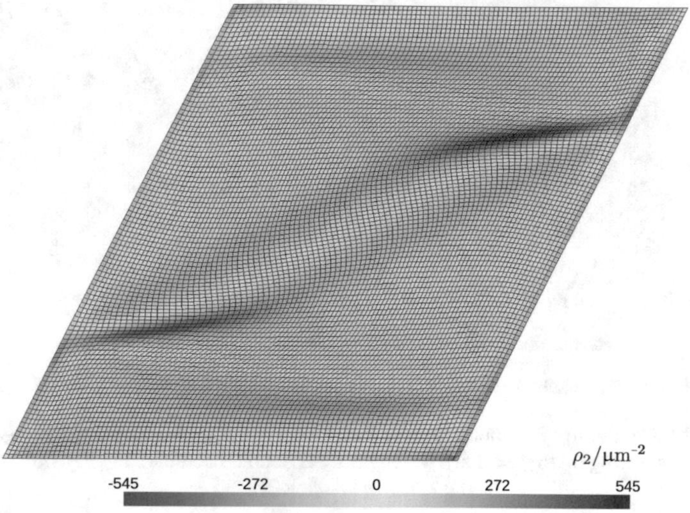

Fig. 7.7 Shear with only SS 2 being active, initial orientation $\varphi = -60°$: Depicted is the signed dislocation density in SS 2 after shearing up to $\gamma = 0.5$. Instead of sharp boundaries, gradual deformation boundaries emerge at $k = 10^2$

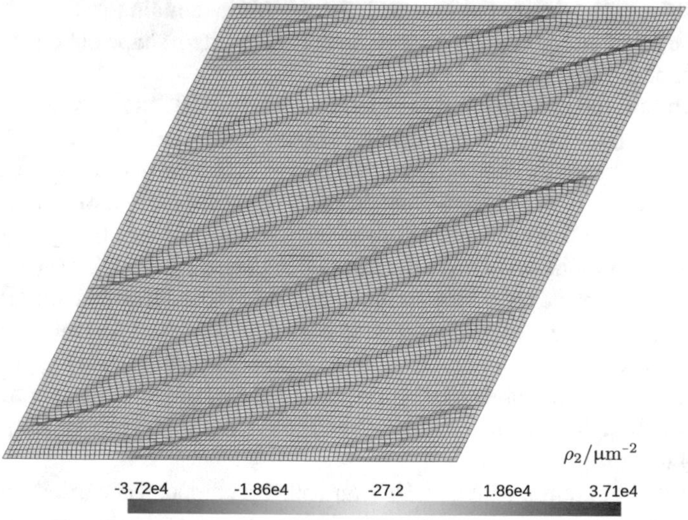

Fig. 7.8 Shear with only SS 2 being active, initial orientation $\varphi = -60°$: Depicted is the signed dislocation density in SS 2 after shearing up to $\gamma = 0.5$. The area is 100 times smaller compared to Fig. 7.6

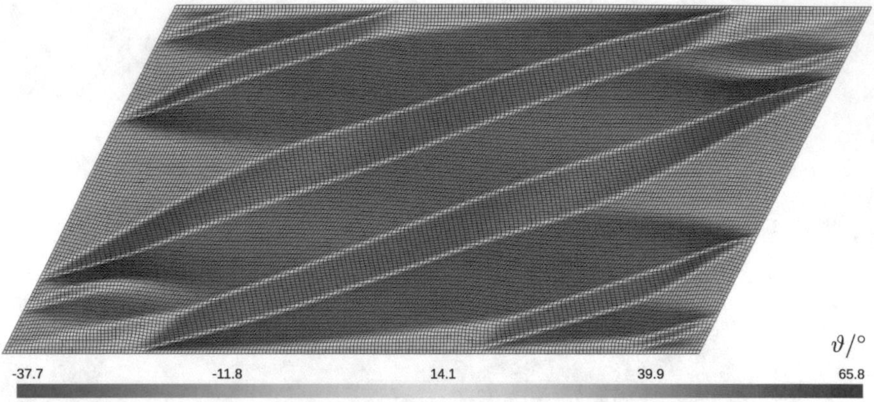

Fig. 7.9 Shear with only SS 2 being active, initial orientation $\varphi = -60°$: Depicted is the lattice rotation after shearing up to $\gamma = 0.5$. The length L_x is doubled compared to Fig. 7.6

In order to estimate the influence of the boundaries on the deformation band's orientation a rectangular crystal domain with $L_x = 2L$, $L_y = L = 200\ \mu\text{m}$ was considered. The simulation result in Fig. 7.9 shows that the orientation of the bands stays the same compared to the case with the quadratic domain (cf. Fig. 7.5). Consequently, the orientation does not depend on the crystallite's shape but on the intrinsic orientation of the slip systems.

Furthermore, the simulation was repeated with $k = 10^{-4}$ for a vanishing plastic flow resistance ($\eta_\text{p} = 0$). In this limiting case the deformation is free of dissipation and the material behavior becomes reversible (cf. p. 24). This also leads to pattern formation (cf. Fig. 7.10), which shows that this does not necessarily have to be associated with dissipation. In this case it is a conservative transfer of free energy by different energy storage mechanisms. When the external load is removed, the system—similar to an elastically buckled rod—immediately returns to its undeformed initial state. The effect and significance of dissipation on this form of structure formation is inferred from this: It leads to *persistent* patterns and substructures and lends the process irreversibility. However, too high dissipation can suppress structure formation. Between the extremes (total loss or total storage of the slip system work) there is the physically relevant and interesting range.

Finally, the simulation with $k = 10^{-4}$ and other parameters according to Table 7.1 is considered once again to investigate the influence of rate dependence on the solution. The whole process is divided into a deformation phase and a relaxation phase. For both phases a duration of $T = 10^{-3}$s is chosen and divided into $N_t = 300$ time steps. Figure 7.11 shows the lattice rotation at the end of the deformation and the relaxation phase. Compared to the simulation with $T = 10^{-5}$s (cf. Figs. 7.5 and 7.6) the following becomes clear: Instead of six areas with negative lattice rotation only two have formed. These are clearly larger and almost rhombic. Due to the 100 times lower deformation rate, correspondingly more plastic slip in SS 2 is possible. This results in an even more regular, regionally homogeneous structure with fewer

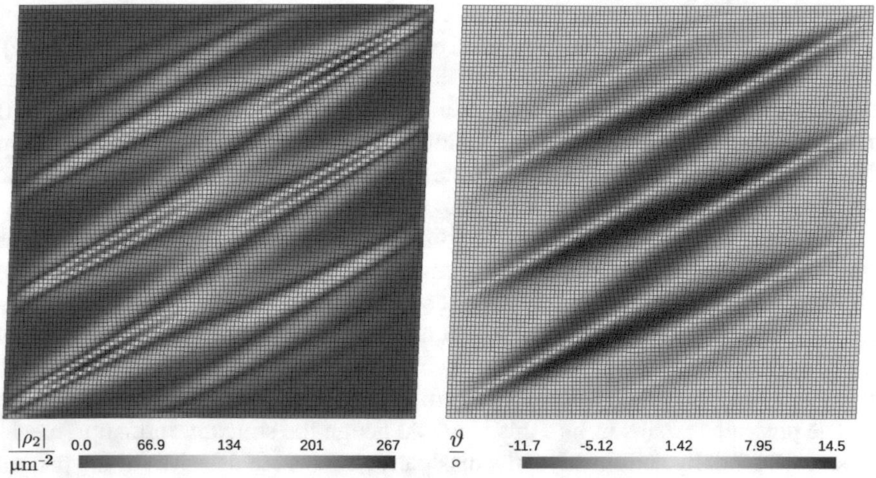

Fig. 7.10 Shear up to $\gamma = 0.1$ with only SS 2 being active, initial orientation $\varphi = -60°$, $\eta_p = 0$: Even at dissipation-free slippage, the pattern-forming instability is evident. *Left* Magnitude of the dislocation density in SS 2; *Right* Lattice rotation

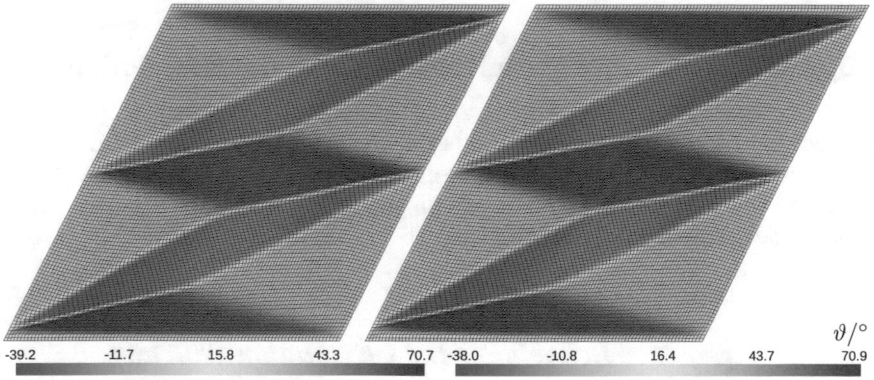

Fig. 7.11 Process with shearing up to $\gamma = 0.5$ and subsequent relaxation, with single-slip in SS 2, initial orientation $\varphi = -60°$: Depicted is the lattice rotation; *Left* at $\gamma = 0.5$ and $t = 10^{-3}$s; *Right* at $\gamma = 0.5$ and $t = 2 \cdot 10^{-3}$s (relaxed state). Corresponding animated graphics can be found at Supplementary material 2 and Supplementary material 3

deformation boundaries. It is also noticeable that there is almost no change in the subsequent relaxation phase. Accordingly, the state after shearing up to $\gamma = 0.5$ at time $t = 10^{-3}$s is already very close to the thermodynamic equilibrium.

For the further classification of these results some aspects from the theory of complex systems are recalled: It is known that a system of many simultaneously interacting, equivalent components under an undirected external influence—which drives the system away from equilibrium—can automatically form a structure [11, 12]. During this self-organization new properties emerge. There are often different mechanisms (stabilizing/destabilizing) which balance each other at critical points (coincidence). Often there is a symmetry breaking instability and the system has to decide between equivalent energy minima or combine them according to the boundary conditions. If a threshold value of the external constraint (e. g. of the deformation) is exceeded at this critical point, a branching to a new state occurs [13, p. 294]. Despite all generic similarities of self-organized systems, the specifics of the system under consideration must always be accounted for [13, p. 290]. Nicolis and Prigogine predicted as early as 1987 that structure formation phenomena in plastic processes should lead to (dynamic) instabilities that induce spatial structures and transitions between multiple stationary states with respect to defect concentration. In the case of supercritical loads, the material re-organizes itself. Large elastic-plastic deformations can provide here for the mentioned distance from the thermodynamic equilibrium.

The presented results fit into this view. Although the structure formation mechanism is essentially conservative, the dissipative effects are decisive for the pattern that really emerges and develops. Accordingly, the subgrain formation obtained is not a "freezing" self-organization process. Dissipation by plastic slip causes entropy outflow and thus enables pattern formation in the form of an ordered state.[3]

7.5 Convergence and Mesh Dependence of the FE Solution

In order to have not only a *qualitative* but also a *quantitative* value, a FE solution should converge to the (unknown) true solution as the element size is successively decreased (*h-convergence*) and/or as the polynomial degree of the shape functions is successively increased (*p-convergence*).

7.5.1 Convergence Behavior of the FE Solution With and Without Localization

It is well-known that FE solutions for non-convex energies and localization phenomena can strongly depend on the underlying FE mesh [14, 15, p. 2]. Therefore, the mesh sensitivity of the obtained results is analyzed in a systematic way in this section. For this, the domain is discretized with different numbers of finite elements.

As a reference, a solution without localization is considered first. Choosing the initial crystal orientation $\varphi = 0$ makes the free energy a convex function. Hence, no pattern formation can occur. The main characteristic of the solution is a pronounced gradient of the primary fields near the boundaries (cf. Fig. 7.12) and GNDs pile up there. Inside the domain there are no significant heterogeneities. The solution is

[3] Even without the thermal problem being considered, the necessary entropy outflow exists by the implicit assumption that all dissipated energy is transferred via heat flow.

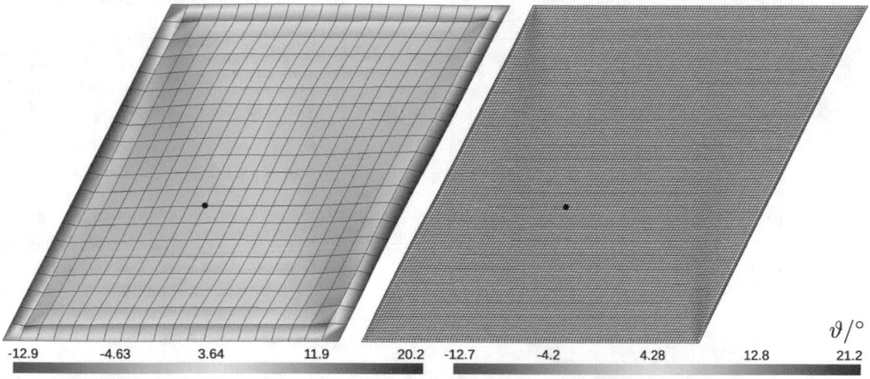

Fig. 7.12 Shear with only SS 2 being active, initial orientation $\varphi = 0°$: Depicted is the lattice rotation after shearing up to $\gamma = 0.5$; *Left* $N_e = 20^2$; *Right* $N_e = 120^2$. Apart from the boundary layer the solution is rather homogeneous. The black dot indicates the location where the evolution of the primary fields is observed

similar to the geometrically linear case [7]. Even though the width of the boundary layer significantly decreases comparing the simulations with $N_e = 20^2$ and $N_e = 120^2$, the minimum and maximum values of the lattice rotation hardly change. Now for the study of the convergence of some FE solution, local quantities are more suitable than integral ones (e. g. reaction forces [14]). The reason is that the latter are less critical in the sense that they might seem identical from one solution to another even though there are still pronounced local differences. Consequently, in order to further study the convergence behavior of the solution, the evolution of the primary field variables u_x, u_y, ν_2 is observed at specific points within the domain. For these observation points the evolution of the relative error between the coarser mesh and the finer mesh was evaluated. The successive reduction of the relative error with decreasing element size is a strong indicator for a converged solution.

Obviously, the difference between the curves decreases with increasing the number of elements, confirming the convergence of the solution (cf. Fig. 7.13). Moreover, the magnitude of the vertical displacement $u_y(t)$ becomes increasingly smaller for higher N_e. This is plausible, as the prescribed shear deformation has no such displacement component either. In addition, it can be observed that the plastic slip rate increases rapidly at the beginning of shearing. Having reached some plateau-like value, there is only a slight oscillation around it for the rest of the deformation phase.

Now a somewhat different situation is considered returning to the case $\varphi = -60°$ with a non-convex free energy and localization. Appearantly, the solution is converged with respect to the dimensions of the bands as, e. g., visible in the lattice rotation field (cf. Fig. 7.14).

However, as it turns out h-convergence is not achieved yet. Refining the mesh produces more and more deformation bands, as becomes particularly clear by increasing the number of elements and the polynomial degree p at the same time (cf. Fig. 7.16). Even within some interval where the number of bands stays constant

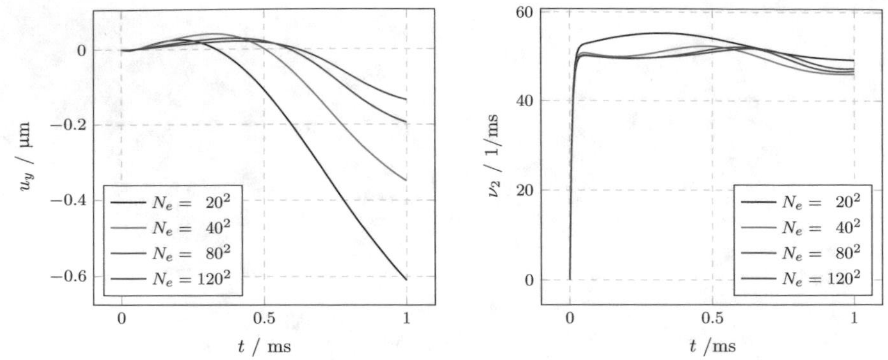

Fig. 7.13 Shear with only SS 2 being active, initial orientation $\varphi = 0°$: Depicted is the evolution of the displacement $u_y(t)$ and the slip rate $\nu_2(t)$ at the location $x = y = 0.4\,L$ for different mesh sizes

(e. g. for $N_e \in [80^2, 110^2]$, cf. Table 7.2), the values of the primary fields in the bands have not fully converged yet. Especially, it was found that it depends on the position of the observation point, whether or not the solution appears converged: For observation points "deep inside" some evolving subgrain (e. g. at $x = y = 0.8\,L$) the relative error between the slip rate curves clearly reduces increasingly with finer mesh (cf. Fig. 7.15, left).

However, for observation points close to or within the evolving subgrain boundaries (e. g. at $x = y = 0.9\,L$), this cannot be claimed for $\nu_2(t)$ (cf. Fig. 7.15, right). From the moment the homogeneous solution becomes instable (cf. also Fig. 7.4), the slip rate curves deviate from each other, ending at different values. Furthermore, the extremal values of the curves differ. However, this observed numerical behavior seems to be in the nature of localized solutions. As the subgrain boundary width de- or increases the character of the local solution at the observation point can drastically change. In this case a comparison is simply not meaningful.

For pure crystal plasticity without any energetical contribution from GNDs, the sharpness or width of the deformation boundaries (subgrain boundaries) would tend to zero with increasingly fine mesh, as there is no intrinsic length scale controlling it [14]. In the continuum dislocation theory, the dislocation energy (with the internal length scale ℓ) plays the role of a regularization. In fact, incorporating GNDs by the dislocation density tensor and the corresponding GND energy with $\ell = 1/(b\rho_s)$ penalizes arbitrarily small oscillations [15, p. 52].

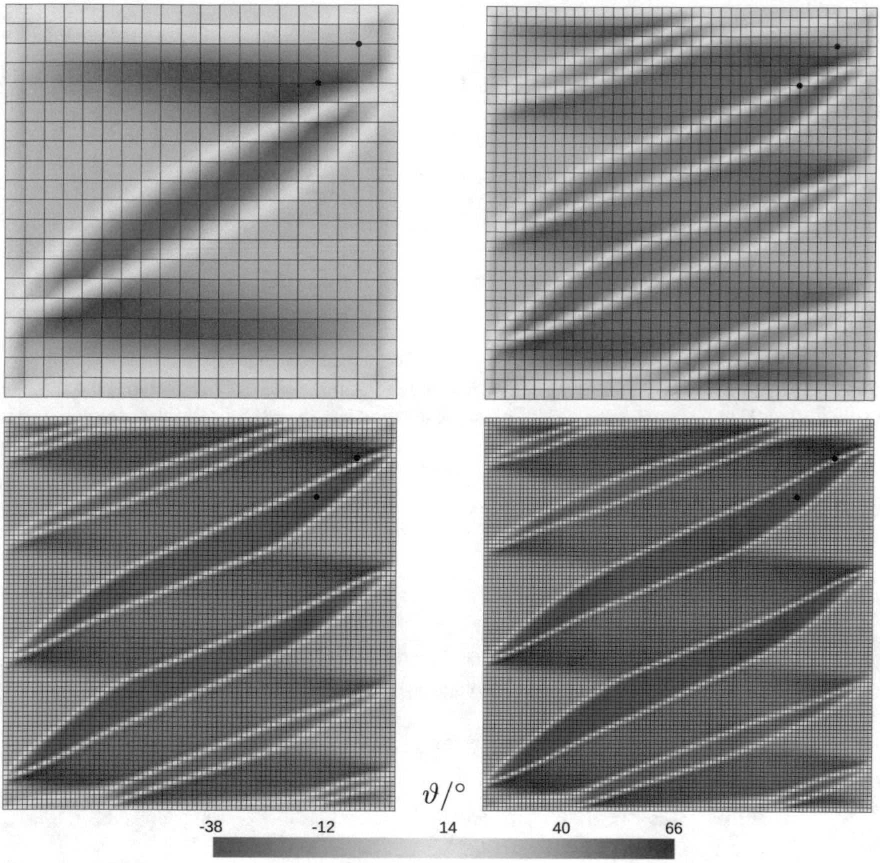

Fig. 7.14 Shear with only SS 2 being active, initial orientation $\varphi = -60°$: Depicted is the lattice rotation on the *undeformed* domain for different mesh sizes; *Top* $N_e = 20^2$, $N_e = 40^2$; *Bottom* $N_e = 80^2$ and $N_e = 100^2$. The black dots indicate the observation points at $x = y = 0.8\,L$ and $x = y = 0.9\,L$

Table 7.2 Number of deformation bands depending on the mesh size

N_e	$10 \cdot 10$	$20 \cdot 20$	$30 \cdot 30$	$40 \cdot 40$	$60 \cdot 60$	$80 \cdot 80$	$100 \cdot 100$	$110 \cdot 110$
Bands	1	2	4	4	4	6	6	6

All in all, the following is stated: Even though the presented FE solutions can depend on the discretization (mesh dependence), the characteristics are always identical: an inhomogeneous deformation is composed of *regionally* homogeneous states. In between, GNDs accumulate and form deformation boundaries. Irrespective of the indicated mathematical challenges, it is concluded that the generic features of localization and pattern (subgrain) formation can be described very well with the presented

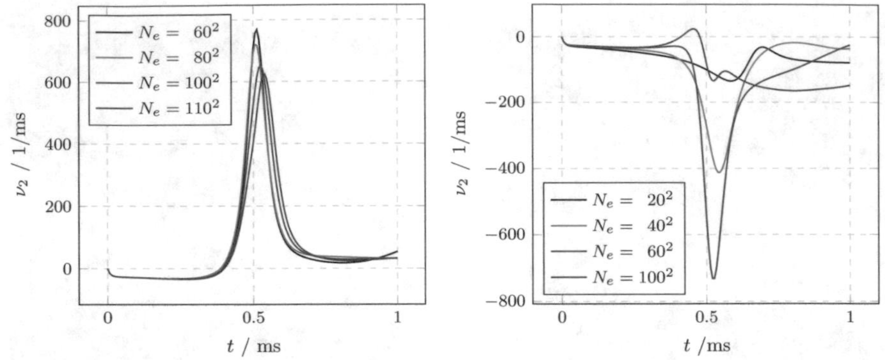

Fig. 7.15 Shear with only SS 2 being active, initial orientation $\varphi = -60°$: Depicted is the slip rate evolution $\nu_2(t)$ for different mesh sizes; *Left* at the location $x = y = 0.8\,L$, *Right* at $x = y = 0.9\,L$

CDT. Still, in future research several open questions have to be addressed and further examined:

1. Does a (unique) solution of the given initial and boundary value problem exist?
2. Under which situations/conditions is the mathematical system well-posed or ill-posed?
3. What is the quantitative link between the width of subgrain boundaries, the internal length scale ℓ and the mesh size?
4. Which material parameters or geometric conditions do control the spacing between the deformation bands?

As has been shown in other studies (cf.e. g. [14, 16]) these questions can be very difficult to answer and lie beyond the scope of the present book.

7.5.2 Possibilities of Improving the Convergence Behavior

In order to improve the convergence behavior of the numerical solution, different steps can be taken: For a robust and accurate Newton-Raphson scheme, the analytical tangent matrix should be used. However, dealing with crystal plasticity calculating the (element) stiffness matrix for implicit FE schemes has generally to be done numerically as a closed analytical form only exists in special cases [17, p. 110]. Another influencing factor of the overall convergence lies in the global FE integration algorithm (cf. Sect. 6.4). Using the power laws (3.37) for the flow rules yields a highly nonlinear relation between the resolved shear stresses and the slip rates. Hence, small variations in the stress state can cause huge variations in the slip rates [17, p. 110].

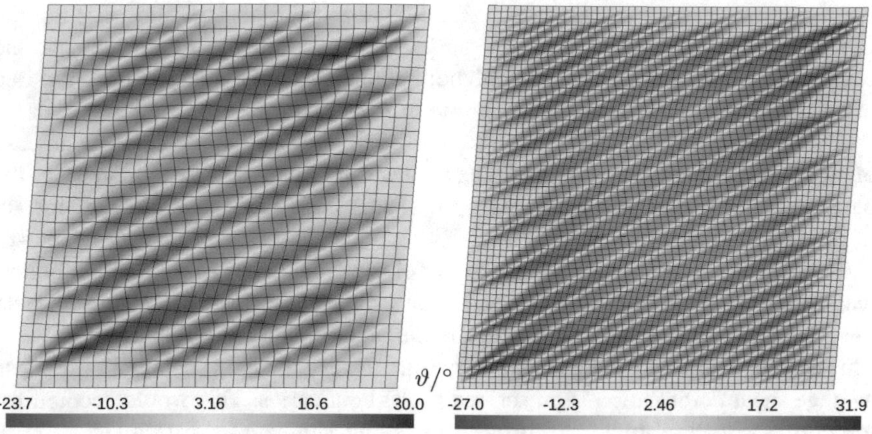

Fig. 7.16 Shear with only SS 2 being active, initial orientation $\varphi = -60°$: Depicted is the lattice rotation for biquadratic shape functions (p-FEM with $p = 2$); *Left* with 30 elements per edge and *Right* with 60

This is effecting up to the calculation of the element and global stiffness matrices and can deteriorate their condition.

Another option to increase the accuracy of the FE solution is to use higher order elements for the following reason: Linear shape functions cannot describe significant gradients of the deformation within one finite element. Consequently, the FE mesh needs to be fine enough to resolve these gradients element by element. In the case of large deformation (i. e. strain and rotation) gradients this is usually not sufficient [17, p. 111 f.]. Then, one possibility is to use as shape functions the integrated Legendre polynomials, which represent a hierarchical functional basis. This has the advantage that the polynomial degree of the shape functions can be increased (so-called p-FEM) to improve the approximate solution without changing the mesh.[4] Applying the polynomial degree 2 in the present simulation example it is found that the number of deformation bands and boundaries considerably increases compared to bilinear elements (cf. Figs. 7.16 and 7.14). Now, a precise laminate structure can emerge as more DOFs are available thus making this mode of deformation possible.

Considering the polynomial degree of the shape functions there is another peculiarity: In the case of mixed finite elements and multi-field problems, the convergence behavior of the solution depends on the (different) choice of the shape functions. In order to obtain optimal convergence the finite element should be constructed in such a way that the so-called infimum-supremum condition is fulfilled [19].[5] The fulfillment of this condition depends in general on the finite element, the mesh topology and the BCs [20, p. 354 ff.]. However, even without evaluating such an inf-sup con-

[4]The associated finite elements are called superparametric [18]. If the polynomial degree equals 1, the well-known bilinear functions result.

[5]Note that this condition is sufficient but not necessary for a reliable and effective finite element scheme [19, 20].

dition, looking at the basic kinematic equation (2.8) suggests the following: For the CDT with displacements $\underline{u}$ and slip rates ν as nodal DOFs it might improve the convergence behavior chosing the polynomial degree for the interpolation of $\underline{u}$ one order higher than that for the interpolation of ν.

Furthermore, using adaptive time stepping can improve the performance in case of slow convergence. However, in doing so it was observed that the symmetry of the band pattern got lost when the time step was drastically reduced at a certain instant of time during the deformation.

Dealing with large deformations, severe element distortions can result in numerical problems. A closer look at Fig. 7.5 shows already considerable distortions near the boundaries, which is attributed to the Dirichlet BCs for the displacement field. This can be significantly relaxed by switching to periodic BCs such that only some average shear deformation is prescribed at the boundaries. This would account for the crystal's "natural trend" to form kinks during laminate formation [15, p. 214]. Likewise, the BCs for the slip rates could be relaxed. Dirichlet BCs for zero slip at the crystallite/grain boundary might be much to strong as they impede any plastic distortion there and thus provoke (unrealistically) high stresses. Instead, the natural BCs (5.14c) could be applied prescribing a value of the micro traction π fulfilling two requirements: (i) it must be sufficiently high to mimic a strong repulsion of GNDs from the grain boundary, and (ii) it must not be so high that the boundary becomes impermeable again.

Instead of solving the FE-discretized variational problem in the straightforward way presented here, there is an alternative way involving quasiconvexity and the theory of relaxation: The minimization of the so-called relaxed variational problem reduces the mesh-dependence of the solution [15, p. 2]. The "lack of convexity of energy densities in variational principles may cause oscillations of minimizing sequences" [15, p. 38], which leads to hierarchical microstructure formation. Hence, analytical and numerical insights obtained from this approach should be taken into account simulating pattern and laminate formation with CDT and FEM.

References

1. Koster, M., Le, K.C.: Formation of grains and dislocation structure of geometrically necessary boundaries. Mater. Sci. Eng. A **643**, 12–16 (2015)
2. Koster, M., Le, K.C., Nguyen, B.D.: Formation of grain boundaries in ductile single crystals at finite plastic deformations. Int. J. Plast. **69**, 134–151 (2015)
3. Silbermann, C.B., Shutov, A.V., Ihlemann, J.: Modeling the evolution of dislocation populations under non-proportional loading. Int. J. Plast. **55**, 58–79 (2014)
4. Hughes, D.A., Hansen, N.: High angle boundaries formed by grain subdivision mechanisms. Acta Materialia **45**(9), 3871–3886 (1997)
5. Zahn, D., Tlatlik, H., Raabe, D.: Modeling of dislocation patterns of small- and high-angle grain boundaries in aluminum. Comput. Mater. Sci. **46**(2), 293–296 (2009)
6. Shutov, A.V., Kuprin, C., Ihlemann, J., Wagner, M.F.X., Silbermann, C.: Experimentelle Untersuchung und numerische Simulation des inkrementellen Umformverhaltens von Stahl 42CrMo4. Materialwiss. Werkstofftech. **41**(9), 765–775 (2010)

7. Silbermann, C.B., Ihlemann, J.: Geometrically linear continuum theory of dislocations revisited from a thermodynamical perspective. Arch. Appl. Mech. **88**(1–2), 141–173 (2017)
8. Läpple, V.: Werkstofftechnik Maschinenbau : theoretische Grundlagen und praktische Anwendungen. Verl. Europa-Lehrmittel, Haan Gruiten (2013)
9. Hull, D., Bacon, D.J.: Introduction to Dislocations, 5th edn. Butterworth-Heinemann, Amsterdam (2011)
10. Sangid, M.D.: The physics of fatigue crack initiation. Int. J. Fatigue **57**, 58–72 (2013)
11. Ebeling, W.: Chaos, Ordnung, Information: Selbstorganisation in Natur und Technik, vol. 74. Harri Deutsch Verlag (1989)
12. Haken, H.: Die Selbststrukturierung der Materie: Synergetik in der unbelebten Welt, German edn. Vieweg, Wiesbaden (1990)
13. Nicolis, G., Prigogine, I., Rebhan, E.: Die Erforschung des Komplexen : Auf dem Weg zu einem neuen Verständnis der Naturwissenschaften. Piper, München, Zürich (1987)
14. Mánica, M.A., Gens, A., Vaunat, J., Ruiz, D.F.: Nonlocal plasticity modelling of strain localisation in stiff clays. Comput. Geotech. **103**(July), 138–150 (2018)
15. Conti, S., Hackl, K. (eds.): Analysis and Computation of Microstructure in Finite Plasticity. Springer, Berlin (2015)
16. Horn, T., Silbermann, C.B., Frint, P., Wagner, M.F.X., Ihlemann, J.: Strain localization during equal-channel angular pressing analyzed by finite element simulations. Metals **8**(1), 55+ (2018)
17. Roters, F., Eisenlohr, P., Bieler, T.R., Raabe, D.: Crystal Plasticity Finite Element Methods in Materials Science and Engineering. Wiley, New York (2010)
18. Szabó, B.A., Babuska, I.: Finite Element Analysis. Wiley, New York (2008)
19. Bathe, K.J.: The inf–sup condition andits evaluation for mixed finite element methods. Comput. Struct. **79**(2), 243–252 (2001)
20. Bathe, K.J.: Finite-Elemente-Methoden. Springer, Berlin (2002)

Chapter 8
Outlook

Abstract This chapter first summarizes the content of the book and discusses key results. Subsequently, a multitude of promising future branches of research is outlined. On the one hand, this concerns possible theoretical extensions of the model and the numerical studies. On the other hand, proper experiments are suggested for the validation of the numerical results. Thereby, the focus is on cubic minerals.

8.1 Summary and Discussion

The geometrically nonlinear theory of dislocation-based crystal plasticity [1], adapted and further specified in the context of this book, is characterized by a consistent coupling of elastic and plastic anisotropy. It takes into account the storage of elastic energy due to lattice distortions and incompatibilities (GNDs) as well as the dissipation of energy during plastic slippage. Therefore, it is considered a continuum dislocation theory (CDT).

From the nonlinear CDT, a mechanical model was derived, which is reduced to the most essential: A plane problem with a simple cubic crystal was considered. The crystal lattice is thus described by two lattice vectors and two (initially perpendicular) slip systems exist. This minimal model considers all effects mentioned above and was implemented in an object-oriented FEM research code.

As a starting point for the investigation of subgrain formation existing studies were used [2, 3]. These suggest that the non-convexity of the free energy with respect to the elastic distortion is a necessary criterion for predicting (dislocation) structure formation. The decisive factor for this non-convexity and geometrical softening is the (elastic) lattice rotation [4, p. 43], which is not present in geometrically linear theory. An energetically favorable final state in the form of a laminate structure was constructed without following the path to this state [3]. The subgrain boundaries correspond to the sharp interfaces between the individual laminate layers. In contrast to this, a *dynamic* model with slip system kinematics on rate level was developed in the present book with the help of the principle of virtual power. The plastic slip rates correspond to phase fields in the sense of the phase field method, with which

C. B. Silbermann et al., *Introduction to Geometrically Nonlinear Continuum Dislocation Theory*, SpringerBriefs in Continuum Mechanics,
https://doi.org/10.1007/978-3-030-63696-8_8

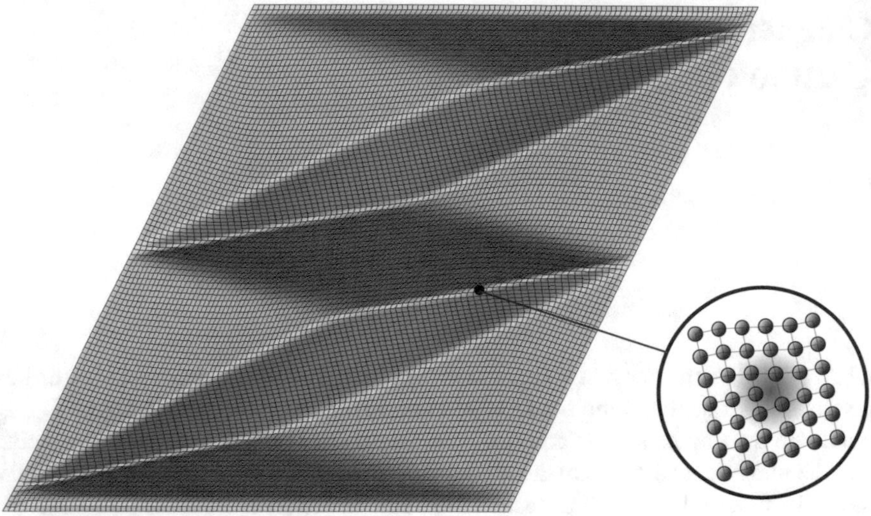

Fig. 8.1 Shear with only SS 2 being active, initial orientation $\varphi = -60°$: Depicted is the lattice rotation according to Fig. 7.1. The tilt boundary is formed by a multitude of edge dislocations, one of which is sketched in the zoom box to the right.

continuous transitions can be modeled instead of sharp interfaces. Thus it was possible to confirm and significantly extend the piecewise homogeneous, static, analytical considerations [2–4] by inhomogeneous, dynamic, numerical simulations. Under single-slip a self-organized (i. e. not constructed) formation and evolution of geometrically necessary dislocation boundaries could be studied under large deformations. The simulated tilt boundaries (cf. Fig. 8.1) formed by GNDs match experimental observations [5]: These defect structures increase proportionally (at the expense of geometrically *redundant* dislocations) and are finally predominant at large deformations. This justifies the direct simulation of subgrain formation by means of a CDT based on GNDs. For the sake of simplicity and comprehensibility, a minimalistic—but large strain—scenario was considered here: a simple cubic single crystal with single slip was loaded in a way which is hardly compatible with the only available plastic deformation mode (slip in system 2). With the crystal's free energy being non-convex and the strains being large enough, this provokes laminate formation in order to fulfill the boundary conditions.

A similar laminate structure formation in single crystals is obtained in 3D simulations with a much more comprehensive model of dislocation-based crystal plasticity [6]. In accordance with the present simulation results, the following is stated: the formation of deformation bands or boundaries requires the (regional) dominance of one slip system; the structure formation process is strongly dependent on the crystallographic orientation; the laminate structure formed as a result of localization is energetically advantageous over homogeneous deformation. In contrast to the present simulation results, the multitude of effects considered renders the structure formation process more complicated, which makes the physical interpretation of the results more difficult.

8.2 Possible Extensions for Future Research

Overall, the geometrically nonlinear CDT and the FE implementation have great potential for future research: In order to further investigate the essential effects of the structure formation process, other elementary tests such as tension/compression and bending can be considered. In doing so, the activity of *both* slip systems can be studied with the presented model. Therefore, several aspects have to be considered. For the correct representation of the real slip behavior, the approach (3.44) with an exponent p as large as possible should be used.[1] Then, the plastic behavior is no longer rate-dependent and differences in the slip systems can manifest themselves clearly (cf. discussion on p. 60). In order to favor the latter, the theory should be adapted in such a way that the true slip rates are primary variables (cf. Remarks 2.1 and 3.3). Thus also the discrepancy outlined in Remark 3.1 disappears.

Furthermore, it can be studied whether and under which parameter constellations the effect of a delocalization [7] can be reproduced with the present model. This is particularly interesting for practical application in forming processes where a homogeneous, stable plastic deformation is desired.

Including the temperature field—as already suggested in Sects. 3.1 and 3.5—the theory can be extended by thermal effects. For this, the kinematics have to take into account thermal expansion, e. g. with a triple multiplicative decomposition of the deformation [8]. Further, the thermal part of the free energy has to be specified and a possible temperature-dependence of mechanical quantities has to be considered too. This concerns thermally activated plastic processes (i. e. dislocation climb) especially, and results in a thermo-mechanical coupling. For the calculation of the temperature field, the first law of thermodynamics has to be evaluated explicitly. Thus, a thermo-viscoplastic model of a continuously dislocated single crystal will be obtained. With the help of this framework, more aspects of microstructural processes can be captured.

In order to also consider geometrically *redundant* dislocations and their interaction with GNDs, the integration of the model of Bortoloni and Cermelli [9] seems particularly worthwhile. On the one hand, the reason for this is that the present work addresses what was neglected there (and vice versa). On the other hand, both models are based on the same theory by Gurtin. In this respect, the approaches complement each other in an optimal way. In addition, such a unified theory could enable the simulation of the successive formation of dislocation cells and subgrains during large deformations. This way, new theoretical insights in the so-called grain refinement could be obtained.

Finally, there is another extension that is compatible with the 2D formulation in the present work. This concerns the transition from single to polycrystals, i. e. the consideration of planar lattice defects in the form of grain boundaries [10]. To this

[1] This requires a solution for the problem $\left.\dfrac{\partial^2 d}{\partial v_i^2}\right|_{v_i=0} \to \infty$ with potential (3.44) at $p > 1$.

end, the crystal domain can be tesselated by Voronoi cells representing the grains, and the grain boundary energy and interaction with dislocations has to be incorporated in the theory.

8.3 Possibilities for Experimental Validation

The numerical simulations of the plane deformation of a single crystal with cubic lattice structure and two active slip systems require an experimental validation. The metal polonium in its cubic primitive α phase would allow a direct comparison, but is practically eliminated due to its radioactivity and associated difficulties.

Among the possible (non-radioactive) monocrystalline cubic materials, ionic crystals with NaCl structure appear particularly suitable. The NaCl lattice structure is cubic (which is why the elastic properties are cubic anisotropic), but not cubic *primitive*. Therefore the slip systems (2.1) have to be adapted. In ionic crystals with NaCl structure the cubic $\{100\}$ slip planes with $\langle 110\rangle$ slip directions and the oblique $\{110\}$ slip planes with $\langle\bar{1}10\rangle$ slip directions are present [11]. Which of these two slip system types is primary or secondary depends on the ionicity of the chemical bond [12]. In the strongly ionic MgO crystals, the critical shear stress τ_{cr} in the oblique slip planes is significantly smaller than in the cubic ones. With decreasing ionicity, e. g. in the series

$$\text{MgO (Magnesia)} \rightarrow \text{NaCl (Halite)} \rightarrow \text{AgCl (Silver halide)} \rightarrow \text{PbS (Galena)},$$

the character of the ionic bond is increasingly attenuated towards the covalent bond, whereby in PbS the cubic slip planes are the primary ones. The choice of the slip planes also depends on the temperature, since the slip of $\{100\}$ planes below $\theta_0^{\{100\}}$ is more difficult. For the plastic deformation of ionic crystals, it is important that the behavior below a temperature θ_{bd} changes from ductile to brittle [12]. If $\theta < \theta_{\mathrm{bd}}$, a cleavage fracture of the ionic crystal may occur before τ_{cr} is reached. For comparison with the present theory, planar slip (of edge dislocations) is also necessary, which often becomes dominant only below a temperature θ_{w}.

Crystals with NaCl structure can be deformed plane, if only the $\{110\}\langle\bar{1}10\rangle$ slip systems are triggered, e. g. for plane deformation in the (001) plane with the orthogonal main slip systems $(110)[\bar{1}10]$ and $(\bar{1}10)[110]$. Under the aforementioned restrictions results for AgCl with $\theta_{\mathrm{bd}} = 80$ K and $\min(\theta_{\mathrm{w}}, \theta_0^{\{100\}}) = 100$ K [12] a temperature range [80K, 69K]. Experiments under these conditions would provide an adequate experimental basis for the verification of this theory. Ionic crystals are also suitable as model material because they may exhibit dislocation structure formation: Low-temperature deformation results in low-energy dislocation configurations with tilt boundaries [13].[2] For comparison with the simulation results, the measurement

[2]The comparison with dislocation cell forming metals and alloys is, of course, limited due to the differences between metallic and ionic bond and the resulting consequences.

of the displacement and deformation field is necessary, for which modern digital image correlation methods [14] could be used.[3]

To ensure that the proposed comparison between experiment and simulation can be done not only *qualitatively* but also *quantitatively*, the material parameters of the model (cf. Table 7.1) must be determined as accurately as possible. Elastic and lattice constants are directly measurable and already tabulated in many cases. Parameters related to the dissipative material behavior (here τ_{cr} and η_{p}) may require an indirect determination. Such a parameter identification could be done by measurements of the dissipated energy [15].

8.4 Concluding Remarks

Having read this book the reader is introducd to single crystal plasticity with continuously distributed dislocations and equipped with the basic ingredients for a FE implementation of this theory. Now, the reader is encouraged to reproduce the presented simulation results, refine the methods and further continue the research in this interesting field of computational materials science. Frenkel's advice sure is useful here (cf. p. v): While pursuing this particular CDT, insights from other theoretical and experimental approaches on dislocation-based crystal plasticity have to be considered and integrated whenever possible.

References

1. Gurtin, M.E.: A gradient theory of single-crystal viscoplasticity that accounts for geometrically necessary dislocations. J. Mech. Phys. Solids **50**(1), 5–32 (2002)
2. Koster, M., Le, K.C.: Formation of grains and dislocation structure of geometrically necessary boundaries. Mater. Sci. Eng. A **643**, 12–16 (2015)
3. Koster, M., Le, K.C., Nguyen, B.D.: Formation of grain boundaries in ductile single crystals at finite plastic deformations. Int. J. Plast. **69**, 134–151 (2015)
4. Conti, S., Hackl, K. (eds.): Analysis and Computation of Microstructure in Finite Plasticity. Springer, Berlin (2015)
5. Ungár, T., Zehetbauer, M.: Stage IV work hardening in cell forming materials, part II: A new mechanism. Scripta Materialia **35**(12), 1467–1473 (1996)
6. Wang, D., Diehl, M., Roters, F., Raabe, D.: On the role of the collinear dislocation interaction in deformation patterning and laminate formation in single crystal plasticity. Mech. Mater. **125**, 70–79 (2018)
7. Yuan, F., Yan, D., Sun, J., Zhou, L., Zhu, Y., Wu, X.: Ductility by shear band delocalization in the nano-layer of gradient structure. Mater. Res. Lett. **7**(1), 12–17 (2018)
8. Shutov, A.V., Ihlemann, J.: Zur Simulation plastischer Umformvorgänge unter Berücksichtigung thermischer Effekte. Materialwissenschaft und Werkstofftechnik **42**(7), 632–638 (2011)
9. Bortoloni, L., Cermelli, P.: Dislocation Patterns and Work-Hardening in Crystalline Plasticity. J. Elast. **76**(2), 113–138 (2004)

[3]The actual realization at low temperatures with optical accessibility of the sample poses a technical challenge.

10. Roters, F., Eisenlohr, P., Hantcherli, L., Tjahjanto, D.D., Bieler, T.R., Raabe, D.: Overview of constitutive laws, kinematics, homogenization and multiscale methods in crystal plasticity finite-element modeling: Theory, experiments, applications. Acta Materialia **58**(4), 1152–1211 (2010)
11. Skrotzki, W., Suzuki, T.: Peierls stresses of ionic crystals with the NaCl-structure. Radiat. Effects **74**(1–4), 315–322 (2006)
12. Skrotzki, W., Frommeyer, O., Haasen, P.: Plasticity of polycrystalline ionic solids. Physica Status Solidi Appl. Res. **66**, 219–228 (1981)
13. Haasen, P., Messerschmidt, U., Skrotzki, W.: Low energy dislocation structures in ionic crystals and semiconductors. Mater. Sci. Eng. **81**, 493–507 (1986)
14. Wang, X.G., Witz, J.F., El Bartali, A., Oudriss, A., Seghir, R., Dufrénoy, P., Feaugas, X., Charkaluk, E.: A dedicated DIC methodology for characterizing plastic deformation in single crystals. Exp. Mech. **56**(7), 1155–1167 (2016)
15. Mróz, Z., Oliferuk, W.: Energy balance and identification of hardening moduli in plastic deformation processes. Int. J. Plast. **18**(3), 379–397 (2002)

Appendix

A.1 Elements of Tensor Calculus and Tensor Analysis

A comprehensive overview of the tensor calculus and analysis used in the present book is given in [1]. This section recalls some elements which are particularly important.

A.1 Isotropic Tensors of Second and Third Order

The isotropic second-order tensor, also known as the metric or identity tensor, reads with respect to some Cartesian base $\underline{e}_a = \{\underline{e}_x, \underline{e}_y, \underline{e}_z\}$:

$$\underline{\underline{I}} = \delta_{ab}\, \underline{e}_a \otimes \underline{e}_b = \underline{e}_a \otimes \underline{e}_a \quad \text{with} \quad \delta_{ab} = \underline{e}_a \cdot \underline{e}_b \,. \tag{A.1}$$

The isotropic third-order tensor, a. k. a. Levi-Cività or Ricci-Permutation tensor reads likewise:

$$\underline{\underline{\underline{\epsilon}}} = -\underline{\underline{I}} \times \underline{\underline{I}} = \underline{e}_a \otimes \underline{e}_b \times \underline{e}_a \otimes \underline{e}_b \quad \text{with} \quad \epsilon_{abc} = (\underline{e}_a \times \underline{e}_b) \cdot \underline{e}_c \,. \tag{A.2}$$

This tensor is fully skew-symmetric, i. e. $\underline{\underline{\underline{\epsilon}}}^{\mathrm{T}} = -\underline{\underline{\underline{\epsilon}}}$ and has the property

$$\underline{\underline{v}} \cdot \underline{\underline{\underline{\epsilon}}}^{\mathrm{T}} \cdot \underline{\underline{w}} = -\underline{\underline{v}} \cdot \underline{\underline{\underline{\epsilon}}} \cdot \underline{\underline{w}} = \underline{\underline{v}} \times \underline{\underline{w}} \qquad \forall\, \underline{\underline{v}}, \underline{\underline{w}} \,. \tag{A.3}$$

Accordingly, $\underline{\underline{\underline{\epsilon}}}$ can be used to convert the cross product between two tensors of any order into a dot product. With $I_3(*)$ denoting the third principal invariant it holds [4]

$$(\underline{\underline{w}} \times \underline{v}) \cdot \underline{\underline{w}}^{\mathrm{T}} = \underline{\underline{w}} \cdot (\underline{\underline{\underline{\epsilon}}} \cdot \underline{v}) \cdot \underline{\underline{w}}^{\mathrm{T}} = I_3\big(\underline{\underline{w}}\big)\, \underline{\underline{\underline{\epsilon}}} \cdot (\underline{\underline{w}}^{-\mathrm{T}} \cdot \underline{v}) \qquad \forall\, \underline{v}, \underline{\underline{w}} \,. \tag{A.4}$$

C. B. Silbermann et al., *Introduction to Geometrically Nonlinear Continuum Dislocation Theory*, SpringerBriefs in Continuum Mechanics,
https://doi.org/10.1007/978-3-030-63696-8

A.2 *Useful Operators and Identities*

The exchange of the second with the fourth base vector and the subsequent symmetrization with respect to the third and fourth base vector is summarized by the following operator:

$$\left(\underline{a} \otimes \underline{b} \otimes \underline{c} \otimes \underline{d}\right)^{S_{24}} = \tfrac{1}{2}\left(\underline{a} \otimes \underline{d} \otimes \underline{c} \otimes \underline{b} + \underline{a} \otimes \underline{d} \otimes \underline{b} \otimes \underline{c}\right) . \tag{A.5}$$

The following identities are valid for tensors of arbitrary order:

$$\operatorname{curl}(\operatorname{grad}(\underline{\underline{U}})) = \underline{\underline{\underline{0}}} \quad , \quad \operatorname{div}(\operatorname{curl}(\underline{\underline{U}})) = \underline{0} \, , \tag{A.6a}$$

$$\operatorname{Curl}(\operatorname{Grad}(\underline{\underline{U}})) = \underline{\underline{\underline{0}}} \quad , \quad \operatorname{Div}(\operatorname{Curl}(\underline{\underline{U}})) = \underline{0} \, . \tag{A.6b}$$

Gradient fields are always curl-free and curl fields are always divergence-free. Applying the differential operators on products of tensors leads to product rules for differentiation, such as:

$$\operatorname{div}(\nu\, \underline{a}) = (\nu\, \underline{a}) \cdot \underline{\nabla} = \nu\, (\underline{a} \cdot \underline{\nabla}) + \underline{a} \cdot (\underline{\nabla} \nu) \, , \tag{A.7a}$$

$$\operatorname{div}(\underline{a} \cdot \underline{\underline{B}}) = (\underline{a} \cdot \underline{\underline{B}}) \cdot \underline{\nabla} = \underline{a} \cdot (\underline{\underline{B}} \cdot \underline{\nabla}) + \underline{\underline{B}} \cdot\cdot (\underline{\nabla} \otimes \underline{a}) \, . \tag{A.7b}$$

A.2 Solutions and Algorithms for Nonlinear Crystal Plasticity

A.1 *Analytical Solution of a Special Form of the Tensor Exponential*

With the plastic slip increments $\xi_i := \nu_i(t{+}\Delta t)\Delta t$ Eq. (6.32) reads in tensor- and matrix representation with respect to the crystal basis $\{\hat{\underline{a}}_{\mathrm{I}}, \hat{\underline{a}}_{\mathrm{II}}, \hat{\underline{a}}_{\mathrm{III}}\}$:

$$\underline{\underline{F}}_{\mathrm{p}}(t{+}\Delta t) = \exp\left(\left(\xi_1\, \hat{\underline{a}}_{\mathrm{II}} \otimes \hat{\underline{a}}_{\mathrm{I}} + \xi_2\, \hat{\underline{a}}_{\mathrm{I}} \otimes \hat{\underline{a}}_{\mathrm{II}}\right)\right) \cdot \underline{\underline{F}}_{\mathrm{p}}(t) \, , \tag{A.8}$$

$$\left[F^{\mathrm{p}}(t{+}\Delta t)\right] = \exp\left(\begin{bmatrix} 0 & \xi_2 & 0 \\ \xi_1 & 0 & 0 \\ 0 & 0 & 0 \end{bmatrix}\right) \left[F^{\mathrm{p}}(t)\right] . \tag{A.9}$$

For the analytical solution of the tensor or matrix exponentials, first the eigenvalue problem is solved for the argument tensor

$$\hat{\underline{\underline{x}}} = \hat{\underline{\underline{L}}}_{\mathrm{p}}\, \Delta t = \xi_1\, \hat{\underline{a}}_{\mathrm{II}} \otimes \hat{\underline{a}}_{\mathrm{I}} + \xi_2\, \hat{\underline{a}}_{\mathrm{I}} \otimes \hat{\underline{a}}_{\mathrm{II}}\,. \tag{A.10}$$

With the only non-vanishing principal invariant $I_2(\hat{\underline{\underline{x}}}) = -\xi_1\xi_2$, the following characteristic equation and its solution results (with $K \in \{\mathrm{I}, \mathrm{II}, \mathrm{III}\}$ denoting here the principal directions):

$$\hat{x}_K^2 - \xi_1\xi_2 = 0 \quad \rightarrow \quad \hat{x}_{\mathrm{I}} = \sqrt{\xi_1\xi_2},\ \hat{x}_{\mathrm{II}} = 0,\ \hat{x}_{\mathrm{III}} = -\sqrt{\xi_1\xi_2}\,. \tag{A.11}$$

The tensor $\hat{\underline{\underline{x}}}$ has the rank 2 and is diagonalizable, because it has three independent left- and right-hand eigenvectors (EVs). These can be determined by solving the system of equations

$$\hat{\underline{g}}^K \cdot \hat{\underline{\underline{x}}} = \hat{x}_{(K)}\, \hat{\underline{g}}^K \quad , \quad \hat{\underline{\underline{x}}} \cdot \hat{\underline{g}}_K = \hat{x}_{(K)}\, \hat{\underline{g}}_K\,. \tag{A.12}$$

A possible solution for the EVs with the appropriate normalization is given by

$$\hat{\underline{g}}^{\mathrm{I}} = \frac{1}{\sqrt{2}}\left(\hat{\underline{a}}_{\mathrm{I}} + \frac{\sqrt{\xi_1\xi_2}}{\xi_1}\,\hat{\underline{a}}_{\mathrm{II}}\right) \quad , \quad \hat{\underline{g}}^{\mathrm{II}} = \hat{\underline{a}}_{\mathrm{III}} \quad , \quad \hat{\underline{g}}^{\mathrm{III}} = \frac{1}{\sqrt{2}}\left(\hat{\underline{a}}_{\mathrm{I}} - \frac{\sqrt{\xi_1\xi_2}}{\xi_1}\,\hat{\underline{a}}_{\mathrm{II}}\right) \quad , \tag{A.13}$$

$$\hat{\underline{g}}_{\mathrm{I}} = \frac{1}{\sqrt{2}}\left(\hat{\underline{a}}_{\mathrm{I}} + \frac{\sqrt{\xi_1\xi_2}}{\xi_2}\,\hat{\underline{a}}_{\mathrm{II}}\right) \quad , \quad \hat{\underline{g}}_{\mathrm{II}} = \hat{\underline{a}}_{\mathrm{III}} \quad , \quad \hat{\underline{g}}_{\mathrm{III}} = \frac{1}{\sqrt{2}}\left(\hat{\underline{a}}_{\mathrm{I}} - \frac{\sqrt{\xi_1\xi_2}}{\xi_2}\,\hat{\underline{a}}_{\mathrm{II}}\right) \quad . \tag{A.14}$$

With these dual base systems the result for $\hat{\underline{\underline{x}}}$ and the tensor exponential function thereof is:

$$\hat{\underline{\underline{x}}} = \sum_K \hat{x}_K\, \hat{\underline{g}}_K \otimes \hat{\underline{g}}^K \quad , \quad \exp\left(\left(\hat{\underline{\underline{x}}}\right)\right) = \sum_K \exp(\hat{x}_K)\, \hat{\underline{g}}_K \otimes \hat{\underline{g}}^K \tag{A.15}$$

and as analytical solution for the tensor exponential of $\hat{\underline{\underline{x}}}$ (A.10) $\forall\, \xi_1, \xi_2 \in \mathbb{R} \setminus 0$ is thus obtained:

$$\exp\left(\left(\hat{\underline{\underline{x}}}\right)\right) = \exp(\sqrt{\xi_1\xi_2})\, \hat{\underline{g}}_{\mathrm{I}} \otimes \hat{\underline{g}}^{\mathrm{I}} + 1\, \hat{\underline{g}}_{\mathrm{II}} \otimes \hat{\underline{g}}^{\mathrm{II}} + \exp(-\sqrt{\xi_1\xi_2})\, \hat{\underline{g}}_{\mathrm{III}} \otimes \hat{\underline{g}}^{\mathrm{III}}\,, \tag{A.16}$$

$$\begin{aligned}\exp\left(\left(\hat{\underline{\underline{x}}}\right)\right) = \ & \tfrac{1}{2}\left(\exp(\sqrt{\xi_1\xi_2}) + \exp(-\sqrt{\xi_1\xi_2})\right)\left(\hat{\underline{a}}_{\mathrm{I}} \otimes \hat{\underline{a}}_{\mathrm{I}} + \hat{\underline{a}}_{\mathrm{II}} \otimes \hat{\underline{a}}_{\mathrm{II}}\right) + 1\, \hat{\underline{a}}_{\mathrm{III}} \otimes \hat{\underline{a}}_{\mathrm{III}} \\ & + \tfrac{1}{2}\left(\exp(\sqrt{\xi_1\xi_2}) - \exp(-\sqrt{\xi_1\xi_2})\right)\sqrt{\xi_1\xi_2}\left(\tfrac{1}{\xi_1}\, \hat{\underline{a}}_{\mathrm{I}} \otimes \hat{\underline{a}}_{\mathrm{II}} + \tfrac{1}{\xi_2}\, \hat{\underline{a}}_{\mathrm{II}} \otimes \hat{\underline{a}}_{\mathrm{I}}\right).\end{aligned} \tag{A.17}$$

The case $\xi_1 = \xi_2 = 0$ leads to the trivial solution $\exp((\underline{\underline{0}})) = \underline{\underline{I}}$. If one slip increment is zero, then $\underline{\underline{\hat{x}}}^n = \underline{\underline{0}}\, \forall\, n \in \mathbb{N}, n \geq 2$ and the series expansion of the tensor exponential [2, p. 9] yields:

$$\xi_1 = 0: \quad \exp\left((\underline{\underline{\hat{x}}})\right) = \underline{\underline{I}} + \xi_2\, \underline{\hat{a}}_{\mathrm{I}} \otimes \underline{\hat{a}}_{\mathrm{II}} \quad , \quad \xi_2 = 0: \quad \exp\left((\underline{\underline{\hat{x}}})\right) = \underline{\underline{I}} + \xi_1\, \underline{\hat{a}}_{\mathrm{II}} \otimes \underline{\hat{a}}_{\mathrm{I}} \,. \tag{A.18}$$

For the numerical evaluation, the case of conjugated complex eigenvalues must also be handled, which result at $\xi_1\xi_2 < 0$. Exploiting

$$\exp(\pm\sqrt{\xi_1\xi_2}) = \exp(\pm\sqrt{-1}\sqrt{-\xi_1\xi_2}) = \cos(\sqrt{-\xi_1\xi_2}) \pm \mathrm{i}\, \sin(\sqrt{-\xi_1\xi_2}) \tag{A.19}$$

yields a real representation of the analytical solution of the tensor exponential of $\underline{\underline{\hat{x}}}$ at $\xi_1\xi_2 < 0$:

$$\begin{aligned} \exp\left((\underline{\underline{\hat{x}}})\right) = &+ \cos(\sqrt{-\xi_1\xi_2})\, \left(\underline{\hat{a}}_{\mathrm{I}} \otimes \underline{\hat{a}}_{\mathrm{I}} + \underline{\hat{a}}_{\mathrm{II}} \otimes \underline{\hat{a}}_{\mathrm{II}}\right) + 1\, \underline{\hat{a}}_{\mathrm{III}} \otimes \underline{\hat{a}}_{\mathrm{III}} \\ &- \sin(\sqrt{-\xi_1\xi_2})\, \sqrt{-\xi_1\xi_2} \left(\tfrac{1}{\xi_1}\, \underline{\hat{a}}_{\mathrm{I}} \otimes \underline{\hat{a}}_{\mathrm{II}} + \tfrac{1}{\xi_2}\, \underline{\hat{a}}_{\mathrm{II}} \otimes \underline{\hat{a}}_{\mathrm{I}}\right) . \end{aligned} \tag{A.20}$$

The correctness of the solution of $\exp((\underline{\underline{\hat{x}}}))$ can be verified by transforming the differential equation for the present 2D special case into a 2×2 matrix form in crystal coordinates, i. e.

$$\dot{\underline{\underline{F}}}_{\mathrm{P}} \cdot \underline{\underline{F}}_{\mathrm{P}}^{-1} = \hat{\underline{\underline{L}}}_{\mathrm{P}} \quad \rightarrow \quad [\dot{F}_{\mathrm{P}}][F_{\mathrm{P}}]^{-1} = \begin{bmatrix} 0 & \nu_2 \\ \nu_1 & 0 \end{bmatrix} . \tag{A.21}$$

The simple structure allows under the constraint $\det[F_{\mathrm{P}}] = 1$ a coefficient-wise analytical solution. By considering Formula (A.17) and (A.20) respectively, one finally obtains exactly the form (A.8)/(A.9).

For the time integration method (6.36), not only the tensor exponential but also its partial derivatives with respect to the slip increments are required, i. e.

$$\frac{\partial}{\partial \xi_i} \exp\left((\underline{\underline{\hat{x}}}(\xi_1, \xi_2))\right) =: \underline{\underline{\hat{e}}}_{,i} \quad \text{with} \quad \underline{\underline{\hat{e}}} = \exp\left((\underline{\underline{\hat{x}}})\right) . \tag{A.22}$$

Since the constant crystal base is independent of the slip increments, only the coefficients in Formula (A.17) resp. (A.20) are to be differentiated. In the <u>*case* $\xi_1\xi_2 > 0$</u> the result is:

$$\begin{aligned} \underline{\underline{\hat{e}}}_{,1} &= \frac{1}{4}\frac{D\,\xi_2}{\sqrt{\xi_1\xi_2}}\, (\underline{\hat{a}}_{\mathrm{I}} \otimes \underline{\hat{a}}_{\mathrm{I}} + \underline{\hat{a}}_{\mathrm{II}} \otimes \underline{\hat{a}}_{\mathrm{II}}) + \frac{1}{4}\left(A - \frac{D}{\sqrt{\xi_1\xi_2}}\right)\frac{\xi_2}{\xi_1}\, \underline{\hat{a}}_{\mathrm{I}} \otimes \underline{\hat{a}}_{\mathrm{II}} + \frac{1}{4}\left(A + \frac{D}{\sqrt{\xi_1\xi_2}}\right) \underline{\hat{a}}_{\mathrm{II}} \otimes \underline{\hat{a}}_{\mathrm{I}}\,, \\ \underline{\underline{\hat{e}}}_{,2} &= \frac{1}{4}\frac{D\,\xi_1}{\sqrt{\xi_1\xi_2}}\, (\underline{\hat{a}}_{\mathrm{I}} \otimes \underline{\hat{a}}_{\mathrm{I}} + \underline{\hat{a}}_{\mathrm{II}} \otimes \underline{\hat{a}}_{\mathrm{II}}) + \frac{1}{4}\left(A - \frac{D}{\sqrt{\xi_1\xi_2}}\right)\frac{\xi_1}{\xi_2}\, \underline{\hat{a}}_{\mathrm{II}} \otimes \underline{\hat{a}}_{\mathrm{I}} + \frac{1}{4}\left(A + \frac{D}{\sqrt{\xi_1\xi_2}}\right) \underline{\hat{a}}_{\mathrm{I}} \otimes \underline{\hat{a}}_{\mathrm{II}}\,, \\ &\text{with} \quad A = \exp(\sqrt{\xi_1\xi_2}) + \exp(-\sqrt{\xi_1\xi_2}) \quad \text{and} \quad D = \exp(\sqrt{\xi_1\xi_2}) - \exp(-\sqrt{\xi_1\xi_2}) \,. \end{aligned}$$

In the <u>$case\ \xi_1\xi_2 < 0$</u> the tensor exponential derivatives with respect to the slip increments are:

$$\underline{\underline{\hat{e}}}_{,1} = \frac{1}{2}\frac{S\,\xi_2}{\sqrt{-\xi_1\xi_2}}\,(\underline{\hat{a}}_{\mathrm{I}}\otimes\underline{\hat{a}}_{\mathrm{I}}+\underline{\hat{a}}_{\mathrm{II}}\otimes\underline{\hat{a}}_{\mathrm{II}}) + \frac{1}{2}\left(C-\frac{S}{\sqrt{-\xi_1\xi_2}}\right)\frac{\xi_2}{\xi_1}\,\underline{\hat{a}}_{\mathrm{I}}\otimes\underline{\hat{a}}_{\mathrm{II}} + \frac{1}{2}\left(C+\frac{S}{\sqrt{-\xi_1\xi_2}}\right)\underline{\hat{a}}_{\mathrm{II}}\otimes\underline{\hat{a}}_{\mathrm{I}}\,,$$

$$\underline{\underline{\hat{e}}}_{,2} = \frac{1}{2}\frac{S\,\xi_1}{\sqrt{-\xi_1\xi_2}}\,(\underline{\hat{a}}_{\mathrm{I}}\otimes\underline{\hat{a}}_{\mathrm{I}}+\underline{\hat{a}}_{\mathrm{II}}\otimes\underline{\hat{a}}_{\mathrm{II}}) + \frac{1}{2}\left(C-\frac{S}{\sqrt{-\xi_1\xi_2}}\right)\frac{\xi_1}{\xi_2}\,\underline{\hat{a}}_{\mathrm{II}}\otimes\underline{\hat{a}}_{\mathrm{I}} + \frac{1}{2}\left(C+\frac{S}{\sqrt{-\xi_1\xi_2}}\right)\underline{\hat{a}}_{\mathrm{I}}\otimes\underline{\hat{a}}_{\mathrm{II}}\,,$$

$$\text{mit}\quad S=\sin(\text{-}\sqrt{\xi_1\xi_2})\quad\text{and}\quad C=\cos(\text{-}\sqrt{\xi_1\xi_2})\,.$$

If one or both slip increments vanish, the derivatives of the tensor exponential can be obtained from the above formulas by applying the limit value theorems and the rule of de l'Hospital. The result in the <u>$case\ \xi_1\xi_2 = 0$</u> is:

$$\underline{\underline{\hat{e}}}_{,1} = \frac{\xi_2}{2}\,(\underline{\hat{a}}_{\mathrm{I}}\otimes\underline{\hat{a}}_{\mathrm{I}}+\underline{\hat{a}}_{\mathrm{II}}\otimes\underline{\hat{a}}_{\mathrm{II}}) + \frac{\xi_2^2}{6}\,\underline{\hat{a}}_{\mathrm{I}}\otimes\underline{\hat{a}}_{\mathrm{II}} + \underline{\hat{a}}_{\mathrm{II}}\otimes\underline{\hat{a}}_{\mathrm{I}} \quad\rightarrow\quad \xi_2=0:\ \underline{\underline{\hat{e}}}_{,1} = \underline{\hat{a}}_{\mathrm{II}}\otimes\underline{\hat{a}}_{\mathrm{I}}\,,$$

$$\underline{\underline{\hat{e}}}_{,2} = \frac{\xi_1}{2}\,(\underline{\hat{a}}_{\mathrm{I}}\otimes\underline{\hat{a}}_{\mathrm{I}}+\underline{\hat{a}}_{\mathrm{II}}\otimes\underline{\hat{a}}_{\mathrm{II}}) + \frac{\xi_1^2}{6}\,\underline{\hat{a}}_{\mathrm{II}}\otimes\underline{\hat{a}}_{\mathrm{I}} + \underline{\hat{a}}_{\mathrm{I}}\otimes\underline{\hat{a}}_{\mathrm{II}} \quad\rightarrow\quad \xi_1=0:\ \underline{\underline{\hat{e}}}_{,2} = \underline{\hat{a}}_{\mathrm{I}}\otimes\underline{\hat{a}}_{\mathrm{II}}\,.$$

This result holds also for the special cases $\xi_1 = 0$, $\xi_2 \neq 0$; $\xi_1 \neq 0$, $\xi_2 = 0$ and $\xi_1 = \xi_2 = 0$.

A.2 Determination of Slip Increments for Rate-Independent Single Crystal Plasticity

Model (4.2) with the implicit flow rule $|\tau_i(\boldsymbol{\nu})| = \tau_{\mathrm{cr}}$ is considered. The slip rates ν_i must be determined such that the inequality (3.35) is fulfilled in every slip system (SS). In order to solve this nonlinear inequality system, an iterative approach is generally required. The difficulty is that the stress state changes during this iteration, making the set of active slip systems variable. A uniform solution becomes possible by transforming the problem into a system of equations with constraints:

$$g_i(\boldsymbol{\nu}) := |\tau_i(\boldsymbol{\nu})| - \tau_{\mathrm{cr}} = 0 \quad\text{with}\quad \tau_i\,\nu_i \geq 0\ \ \forall\ i\,. \tag{A.23}$$

The solution algorithm is derived w. l. o. g. on the example of plane deformation of a cubic primitive crystal with two slip systems 1 and 2. The nonlinear equation system $g_1(\boldsymbol{\nu}) = 0$, $g_2(\boldsymbol{\nu}) = 0$ is solved with a Newton iteration over $n = 0, 1, 2, \ldots$ steps:

$$\begin{bmatrix} {}^{n+1}\xi_1 - {}^{n}\xi_1 \\ {}^{n+1}\xi_2 - {}^{n}\xi_1 \end{bmatrix} = \begin{bmatrix} \Delta\xi_1 \\ \Delta\xi_2 \end{bmatrix} = -\left.\begin{bmatrix} g_{1,1} & g_{1,2} \\ g_{2,1} & g_{2,2} \end{bmatrix}\right|_{{}^{n}\boldsymbol{\xi}}^{-1} \left.\begin{bmatrix} g_1 \\ g_2 \end{bmatrix}\right|_{{}^{n}\boldsymbol{\xi}} \quad\text{with}\quad {}^{n}\boldsymbol{\xi} = \begin{bmatrix} {}^{n}\nu_1 \\ {}^{n}\nu_2 \end{bmatrix}\Delta t\,. \tag{A.24}$$

Here, the start values ${}^{0}\boldsymbol{\xi} = \mathbf{0}$ (no plastic flow) are useful. The partial derivatives

$$g_{i,k} = \frac{\partial g_i}{\partial \xi_k} = \frac{\partial |\tau_i(\boldsymbol{\xi})|}{\partial \xi_k} = \frac{\text{sign}(\tau_i)}{j} \left\{ \frac{\partial \hat{\underline{\underline{C}}}_{\text{e}}}{\partial \xi_k} \cdot \underline{\underline{\underline{\underline{\widehat{K}}}}} \cdot\cdot \underline{\underline{\hat{E}}}_{\text{e}} + \underline{\underline{\hat{C}}}_{\text{e}} \cdot \underline{\underline{\underline{\underline{\widehat{K}}}}} \cdot\cdot \frac{1}{2} \frac{\partial \hat{\underline{\underline{C}}}_{\text{e}}}{\partial \xi_k} \right\} \cdot\cdot \underline{\underline{\hat{S}}}_i^{\text{T}} \tag{A.25}$$

result according to the definition (3.34) of the shear stress τ_i in SS i, where the volume stretch $j = I_3(\underline{\underline{F}}_{\text{e}})$ does not depend on the plastic slip. Each Newton step n starts with the assumption that all slip systems are active. The solution so obtained from Eq. (A.24) at n+1 is subject to the proviso that plastic slip occurs only as long as $|\tau_i| \geq \tau_{\text{cr}}$ holds.[1] To ensure this, after each iteration step, a check is made whether the consistency condition (3.33) is fulfilled. If not, the corresponding slip system should not have been active and the slip increment is reset:

$$^{n+1}\tau_k \, ^{n+1}\xi_k < 0 \; : \qquad ^{n+1}\xi_k \stackrel{!}{=} 0 \quad \leftrightarrow \quad \Delta\xi_k = {}^{n+1}\xi_k - {}^{n}\xi_k = -{}^{n}\xi_k \, . \tag{A.26}$$

After inserting this "boundary condition", the system of equations is solved again. For this purpose the matrix of the slip increments is divided into a searched (S) and a known (K) part and then the reduced system of equations is evaluated:

$$[\Delta\xi] = [\overset{\text{S}}{\Delta\xi}] + [\overset{\text{K}}{\Delta\xi}] \quad \rightarrow \quad [\overset{\text{S}}{\Delta\xi}] = -[g_{i,k}]^{-1}[g_k] - [\overset{\text{K}}{\Delta\xi}] \, . \tag{A.27}$$

If, e. g., it was found that in iteration n+1 the SS 1 had to be inactive, the following is done:

$$\Delta\xi_1 = -{}^{n}\xi_1 \; : \qquad [\overset{\text{S}}{\Delta\xi}] = \begin{bmatrix} 0 \\ \Delta\xi_2 \end{bmatrix} = -\begin{bmatrix} g_{1,1} & g_{1,2} \\ g_{2,1} & g_{2,2} \end{bmatrix}^{-1}\Bigg|_{^n\xi} \begin{bmatrix} g_1 \\ g_2 \end{bmatrix}\Bigg|_{^n\xi} - \begin{bmatrix} -{}^{n}\xi_1 \\ 0 \end{bmatrix} , \tag{A.28}$$

from which $\Delta\xi_2 = (g_{2,1}\,{}^{n}\xi_1 - g_2)/g_{2,2}$ results. Under the "BC" $\Delta\xi_2 = -{}^{n}\xi_2$ the solution follows analogously as $\Delta\xi_1 = (g_{1,2}\,{}^{n}\xi_2 - g_1)/g_{1,1}$. It is important that the corresponding slip systems can also only be "temporarily" inactive. If the stress state changes in the next iteration step, the set of active slip systems can change accordingly. This iteration is terminated when the norm of the step size $\Delta\boldsymbol{\xi}$ is small enough and $\tau_k\,\xi_k \geq 0$ is fulfilled in all slip systems. In the present example the condition reads $\Delta\xi_1^2 + \Delta\xi_1^2 < \epsilon_{\text{nu}}^2$ and $\tau_1\,\xi_1 \geq 0$, $\tau_2\,\xi_2 \geq 0$.

The described procedure can be understood as a "subspace method" and is similar to the handling of contact problems: A searched displacement field must satisfy boundary conditions in the form of inequalities in order to prevent penetration of the bodies in contact. The algorithm can also be transferred to single crystal plasticity considering GNDs, where τ_i has to be replaced by the *effective* shear stress κ_i according to Formula (3.32). If the unknown slip rates or increments represent (nodal) degrees of freedom, it must be ensured, however, that the functional base

[1] Otherwise, there may be a "backflow" into the elastic region, which is inadmissible here.

can reproduce kinks. If the ξ_i can be treated as inner variables, this problem does not arise (cf. discussion in Sect. 6.4).

Finally, it is emphasized that the algorithm was successfully tested for two slip systems. However, depending on the number of (active) slip systems, ambiguity might arise. In order to still obtain a unique solution, further constraints have to be applied then (cf. [3]).

References

1. Silbermann, C.B., Ihlemann, J.: Geometrically linear continuum theory of dislocations revisited from a thermodynamical perspective. Arch. Appl. Mech. **88**(1–2), 141–173 (2017)
2. Roters, F., Eisenlohr, P., Hantcherli, L., Tjahjanto, D.D., Bieler, T.R., Raabe, D.: Overview of constitutive laws, kinematics, homogenization and multiscale methods in crystal plasticity finite-element modeling: Theory, experiments, applications. Acta Materialia **58**(4), 1152–1211 (2010)
3. Gurtin, M.E.: A gradient theory of single-crystal viscoplasticity that accounts for geometrically necessary dislocations. J. Mech. Phys. Solids **50**(1), 5–32 (2002)
4. Donner, H.: FEM-basierte Modellierung stark anisotroper Hybridcord-Elastomer-Verbunde. Ph.D. thesis, Technische Universität Chemnitz (2017)

Index

C. B. Silbermann et al., *Introduction to Geometrically Nonlinear Continuum Dislocation Theory*, SpringerBriefs in Continuum Mechanics,
https://doi.org/10.1007/978-3-030-63696-8

Printed in the United States
by Baker & Taylor Publisher Services